THE SOUL FALLACY

THE SOUL FALLACY

WHAT SCIENCE SHOWS WE GAIN
FROM LETTING GO OF OUR SOUL BELIEFS

JULIEN MUSOLINO

Prometheus Books

59 John Glenn Drive
Amherst, New York 14228

Published 2015 by Prometheus Books

Cover design by Jacqueline Nasso Cooke
Cover image © chaoss / Media Bakery
Unless otherwise noted, all artwork was created by the author
using images from DepositPhotos.com.

Every attempt has been made to trace accurate ownership of copyrighted material in this book. Errors and omissions will be corrected in subsequent editions, provided that notification is sent to the publisher.

Prometheus Books recognizes the following registered trademarks and trademarks mentioned within the text: Escort™, Ford ®, iMac®, iMovie®, iPhone®, Ouija®, PlayStation®, Super Bowl®, YouTube®.

Inquiries should be addressed to
Prometheus Books
59 John Glenn Drive
Amherst, New York 14228
VOICE: 716–691–0133
FAX: 716–691–0137
WWW.PROMETHEUSBOOKS.COM

19 18 17 16 15   5 4 3 2 1

Library of Congress Cataloging-in-Publication Data Pending

Musolino, Julien.
     The soul fallacy : what science shows we gain from letting go of our soul beliefs / by Julien Musolino.
          pages cm
     Includes bibliographical references and index.
     ISBN 978-1-61614-962-8 (pbk.) — ISBN 978-1-61614-963-5 (ebook)
     1. Soul. 2. Religion and science. I. Title.

BL290.M87 2015
202'.2—dc23

                                                                                2014026944

Printed in the United States of America

CONTENTS

FOREWORD

Victor J. Stenger

Gilgamesh, where are you hurrying to?
You will never find that life for which you are looking.
When the gods created man, they allotted him death,
But life they retained in their own keeping.
Gilgamesh, fill your belly with good things;
Day and night, night and day, dance and be merry, feast and rejoice.
Let your clothes be fresh, bathe yourself in water,
Cherish the little child that holds your hand,
And make your wife happy in your embrace;
For this too is the lot of man.
　　　　　　—*The Epic of Gilgamesh* (**third millennium BCE**)[1]

The common wisdom among scientists and theologians is that science has nothing to say about God. In an official statement, the US National Academy of Sciences, the same organization that shamefully denied membership to Carl Sagan, declared in 1998: "Science is a way of knowing about the natural world. It is limited to explaining the natural world through natural causes. Science can say nothing about the supernatural. Whether God exists or not is a question about which science is neutral."[2] Christian theologian John F. Haught agrees, writing, "The scientific method by definition has nothing to say about God, meaning, values, or purpose."[3]

It is also widely claimed that no conflict exists between religion and science. In their 2009 book *Unscientific America*, journalist Chris Mooney and former congressional science fellow Sheril Kirshenbaum tell us, "A great many religious organizations in America uphold the principle of compatibility" of religion and science.[4]

Apparently these organizations have not informed their members of this policy. A 2009 poll by the Pew Research Center found that 45 percent of Republicans (who are often conservative and Christian) see a conflict between science and their beliefs.[5] As for Democrats, 33 percent also say that science conflicts with their faith. Clearly a large number of Americans—liberal and conservative—believe that science and religion are incompatible.

Carl Sagan is often quoted as having said, "Absence of evidence is not evidence of absence." This kept him on good terms with believers. However, this statement is not always true, as I pointed out in my 2007 book *God: The Failed Hypothesis*.[6] When the evidence that is absent should be there and is not, then that can be taken as evidence of absence.

On winter mornings I can look out my office window on an empty field covered with fresh snow. Occasionally I will see footprints of wild animals—foxes, coyotes, and other unidentified wildlife. I rarely see the animals themselves, but I know they exist from the fact that they left footprints. The God most people worship plays a very active role in the universe and in human lives. There should be ample evidence for his existence. But there is none. God has left no footprints on the snows of time.

For millennia humans assumed that gods and spirits were responsible for the events and phenomena they observed. With dangers all around them, the brains of primitive people evolved by natural selection to ascribe agency and intentionality to every event, which enabled humanity to survive. While this made intuitive sense to our ancestors, some began to doubt that invisible agents whose goals and purposes are modeled after our own govern the world. In the sixth century BCE, Greek philosophers, beginning with Thales of Miletus (ca. 624–546 BCE), proposed purely natural explanations for physical phenomena that involved only inanimate ingredients such as water that could be seen, and forces such as the wind that could be felt. This was the beginning of science.

Much of the scholarship of ancient philosophers and scientists was lost to Europe during the thousand-year era termed the Dark Ages. With the autocratic Catholic Church dominating life in Christendom throughout that period, little scientific progress was made. Fortunately, many ancient texts were saved, translated, and expanded upon by Arabic scholars during the golden age of Islam that began in 622 and ended in 1492—just as that knowledge had begun to creep back into Europe.

Until then, Europeans continued to rely on supernatural explanations for what they did not understand. The complex phenomena they saw in the world were taken as evidence for the existence of an intelligent creator whom they assumed to be the Christian God.

Today this rationale is known as the "god-of-the-gaps" argument, also referred to as the "argument from ignorance." When a person says he or she cannot understand how something, such as the human body or the cosmos, could have come about naturally, he or she concludes that it must have had a supernatural cause. But just because someone—or even everyone at the time—does not understand some phenomenon, that doesn't mean a natural explanation for the phenomenon will never be found. History suggests otherwise.

The Dark Ages ended when the Renaissance and Reformation challenged the authority of the Catholic Church in the sixteenth century. This led to more open thinking and the rise of the new science that provided increasingly precise physical explanations for natural phenomena. As the gaps in our understanding were steadily plugged by science, God was no longer needed to fill them. Today we have Darwinian evolution, which eliminates the need for God to explain the complexity of life. We have several scenarios for a natural origin of the universe that require no creator or act of creation. The germ theory of disease means we no longer have to blame illnesses on the devil and engage in painful exorcisms. We can provide natural explanations for earthquakes, tornadoes, hurricanes, tsunamis, and other destructive events that once were attributed to divine throwers of thunderbolts.

And, as we will see in this beautifully written, impeccably researched, and compassionate book by Rutgers psychologist Julien Musolino, we are now in a position to eliminate perhaps the most deeply personal and destructive superstition held by the bulk of humanity: the immaterial, immortal *soul*.

Musolino provides a compelling scientific case for the nonexistence of the soul. The author makes it clear that he is not presenting his own idiosyncratic view but that of the wide consensus of psychologists, cognitive scientists, and neuroscientists—not to mention biologists and physicists. As a physicist who has never seen the hand of God in any observable phenomenon, I view the scientific rejection of the soul as the ultimate validation of reductionist material monism.

The elimination of the soul will be the final, fatal blow to religious belief. The idea of the duality of mind and body has been taken for granted by most religious believers for centuries. Today we also find mind-body duality to be a common presumption among those who have disassociated themselves from organized religion but have not adopted total materialism, describing themselves as "spiritual but not religious."

No one can deny the obvious connection between mind and the physical brain, and not all of those who call themselves "spiritual" embrace supernaturalism. I wish they would find some other term. The word *spirit* usually refers to some substance other than matter—the stuff of the soul.

No doubt soul belief is widespread in America. A 2013 Harris poll showed that 64 percent of US adults believe in the survival of the soul after death, which is down from 69 percent in 2005 but still substantial.[7] Musolino found that even a large majority of his own students in psychology classes, where the mind is presented as an abstract description of the physical workings of the brain, still believe they have an immaterial soul that will carry their consciousness into the afterlife.

Many of today's science-savvy believers, including a minority of scientists, think they can reconcile their faith with the biological and physical sciences that often seem in conflict with their beliefs: humans evolved, but God guided the process; the universe began 13.8 billion years ago rather than 6,000 years ago, but God still created it; there is no scientific evidence for God, but he has his reasons to hide from us.

However, squaring soul belief with science is going to be considerably more challenging. How can supernatural religion and spirituality exist without the immaterial, immortal soul?

As Musolino explains in clear, nontechnical language, there is no need to introduce a supernatural soul in order to understand human thinking. If our memories and consciousness are to survive death, as most religious people believe, then memory and consciousness must reside in some nonmaterial entity. But if that's the case, then why do we lose memories when we experience brain trauma or disease, or ingest drugs? If consciousness dwells in the soul, then why do we become unconscious under anesthesia or from a blow to the head?

The popular conception of consciousness is one of awareness, intent, and feeling operating on the information provided by our senses and memories. That information provides us with a knowledge base for our actions, which are then performed consciously by an entity called the "self" that is thought to be the essence of our personhood and is usually identified with the soul. For example, when I lift a fork to my lips at the dinner table, my conscious self performs a deliberate act by telling my arm and hand what to do. At least, this is the common understanding of what is called *free will*.

However, laboratory experiments pioneered in the 1980s by physiologist Benjamin Libet have shown that before we become aware of making a decision, our brains have already laid the groundwork for that decision.[8] While the interpretation of Libet's original results remains controversial,[9] continuing research has strongly confirmed the main feature of the phenomenon, which is the significant time delay between the brain beginning to shape a decision and our awareness of making that decision.[10] The fact that we have conscious access to only a fraction of all brain processing should give us pause as we consider decisions, intuitions, or feelings of certainty. The major role of the unconscious in our behavior and decision-making challenges assumptions about free will and the associated religious teachings about sin and redemption, as well as our judicial concepts of responsibility and punishment. If our brains are making our decisions for us outside conscious awareness, how can we be responsible for our actions? How can our legal system punish criminals or God punish sinners who aren't in full control of their decision-making faculties?[11] This book addresses these important questions.

Of course, most people will not immediately welcome the message that the soul does not exist and that there is no life after death. But what should Musolino and his colleagues in psychology, cognitive science, and neuroscience do? Hide the results of their research just to make people feel better? Keep their conclusions confined to the technical literature and not explain them to the general public? Allow the great propaganda machines of organized religion and New Age spirituality to continue to delude people with promises of life after death that are based on ancient superstitions and unsupported by evidence?

Musolino does not leave it at that but suggests, as many atheists in the past have also advocated, that we are far better off knowing that we don't have souls than being trapped in the prison of soul belief. Even if you are a moderate believer who is open to almost everything science provides, the very fact that you still indulge in magical thinking leaves you prone to making decisions in your life based on ideas that have no rational support. For example, you might deny your child medical care or vaccination because to do so would conflict with the teachings of your church.

It will take a long time, but humans are going to have to come to grips with the fact that they have no souls. Believers can't go on living in a fantasy world that does them more harm than good. Even in America, with its exceptionally high religiosity compared to the rest of the developed world, reality must eventually intrude. Some say this will never happen. However, it is happening very dramatically elsewhere. Here we do not have to theorize but simply look at the data, which show that not only is it possible for people to abandon belief, but the result is a far happier and healthier society than that found in God-intoxicated America.

In 2005, freelance researcher Gregory S. Paul published the results of surveys that correlate religiosity in various first-world nations with social health. He concluded that "in almost all regards the highly secular democracies consistently enjoy low rates of social dysfunction, while pro-religious and antievolution America performs poorly."[12] Similar results are reported by sociologists Pippa Norris and Ronald Ingelhart in their 2011 book *Sacred and Secular*[13] and by sociologist Phil Zuckerman in his 2009 book *Society without God.*[14]

The facts are clear and simple: many people are happier and behave well—even better—without belief in God, and there is no reason to think the rest of humanity will never come around and finally discard the supernatural fantasies that endanger the very survival of our species. After all, humans have become accustomed to the fact that they are not at the center of the universe, as well as the fact that they are not unique creations but just another animal that happens to have evolved a highly developed brain.

As mentioned, the brain evolved by natural selection to have certain characteristics that enabled early humans to survive in their hostile environment. That environment no longer exists, but we still possess the same brain because it has not had time to evolve further. We are now in a postevolutionary phase where we must use intellect to overcome our more natural animal instincts.

That intellect has recognized that we all must die someday, but it has also given us the ability to use our reason to accept that fact and live happy, rewarding lives in the finite time we have.

The author of this foreword is grateful for the comments and suggestions provided by Greg Bart, Peter Boghossian, John Crisp, Don McGee, Brent Meeker, Anne O'Reilly, Brian Silston, Phil Thrift, Jim Wyman, and Bob Zannelli.

PREFACE

I have put into thy hands what has been the diversion of some of my idle and heavy hours. If it has the good luck to prove so of any of thine, and thou hast but half so much pleasure in reading as I had in writing it, thou wilt as little think thy money, as I do my pains, ill bestowed.

—**John Locke, "Epistle to the Reader,"**
An Essay Concerning Human Understanding, **1689**

In case you haven't noticed, heaven is in the air. In 2004, a book called *90 Minutes in Heaven,* hit the bestseller list. The author, Baptist minister Don Piper, described how his Ford Escort collided with an eighteen-wheeler, sending him straight to heaven. This title remained on the bestseller list for nearly four years. Two years ago, *Newsweek* magazine announced on its front cover that heaven was real. A former Harvard Medical School neurosurgeon named Eben Alexander had a near-death experience and claimed that he went to heaven. Alexander wrote about his story in a sensational bestseller called *Proof of Heaven.* Last month, Hollywood released a movie called *Heaven Is for Real,* based on the bestselling 2010 book by the same name. Christian authors, New Age gurus, and a handful of professional scientists have declared in a flurry of recent popular books that science itself proves that the immaterial spirit can survive death. The 1990s was the decade of the brain. Welcome to the decades of the soul.

Books touting the immortality of the soul and the reality of heaven do so well here in the United States because they resonate with the cultural zeitgeist. According to national polls, most Americans believe that the soul is real, and a chorus of conservative voices insists that without belief in immortality, life on earth would be devoid of meaning and our society would be doomed to abject failure. I decided to write this book because I believe that the soul narrative is profoundly mistaken. A colleague of mine once quipped that if I wanted to make it big, I should write a book with the word *heaven* in the title. Instead, I chose to write a book called *The Soul Fallacy*. But don't be alarmed by the title. This is a friendly book with a happy ending. It is intended for anyone who has ever wondered about the question of the soul.

If you believe in the soul, I urge you to read this book. I hope that it will change your mind, but even if it doesn't, it will at least shed a truer and kinder light on those of us who do not share your beliefs and are so often vilified in our pro-religious culture. If you do not believe in the soul, this book will be grist for your mill and will show you step by step why your skepticism is both imminently reasonable and practically beneficial. If you are unsure whether people have souls, you will find what you need in this book to make up your mind and join the growing ranks of the soul-free. If you are just curious about the soul, you will discover in the pages ahead where soul beliefs come from, how they changed over the course of history, why they come to us so naturally, how they affect our conceptions of life, justice, and moral responsibility, why we do not need to believe in the immortality of the soul to find meaning and purpose in life, and above all, what mainstream science really says about the soul.

There are four important conclusions that I would like to share with you in this book. The first is that the traditional idea of an immortal soul is as much a scientific hypothesis as it is a metaphysical or religious claim. Consequently, deciding whether we have a soul is an objective endeavor that falls squarely within the scope of scientific inquiry. The second conclusion is that in spite of well-publicized claims to the contrary, there is in fact no credible evidence supporting the existence of the soul. The third conclusion is that modern science gives us every reason to believe that people do not have souls. My final, and perhaps most important conclusion, echoing the words of Charles Darwin, is that there is grandeur in this view of life—and death—and that we lose nothing, morally, spiritually, or aesthetically by giving up our soul beliefs. In fact, I will show you that we even have something to gain. To reach these conclusions, I will take you on a tour of history, science, and philosophy using accessible language and without assuming any specialized knowledge.

The ideas and conclusions presented in this book did not come fully shaped in their final formulation. Like us, they evolved, and they represent the fruit of centuries of hard work, false starts, and insights by individuals whose intellectual prowess has left an indelible mark on the fabric of history. In sharing with you their accumulated knowledge and wisdom, I have chosen to be the chronicler of their ideas, weaving together the ones that I believe will present the most compelling tapestry, a lifelike mural depicting the story of us. Yet this remains a challenging task, and I cannot pretend that I will be able to do justice to every idea, concept, or argument. After all, this is a popular book, not an academic treatise. By necessity, I will also leave out certain ideas, details, observations, and conclusions that some might regard as important. While fully acknowledging these limitations, I also believe that few working scientists who have pondered the same questions today or written about them will disagree with my general conclusions.

Just as great minds of past and current generations need to be acknowledged, so do a number of people, fellow academics and nonscientists, loved ones, and friends and colleagues, who have had a great influence on my own thinking and the creation (or perhaps better, the *evolution*) of this book. I would like to extend my first words of gratitude to my literary agents Joanna Swainson and Gillian MacKenzie. Navigating the world of trade publishing is a challenging task, especially for a new author, and *The Soul Fallacy* would never have seen the light of day without Joanna and Gillian's endorsement, their guidance, and their constant encouragements. I am also grateful to Prometheus Books for adding *The Soul Fallacy* to its list and for its passionate advocacy of free thought, science, and reason over the years. Like the Greek titan it represents, the Prometheus brand remains a stable beacon of light and reason in our demon-haunted world.

A colleague once told me that one of the perks of being an academic is that we are surrounded by smart people. At Rutgers University, and in the academic community at large, I have been surrounded by many smart people who have imparted their knowledge and wisdom to me and with whom I have had the privilege to discuss the ideas presented in this book. Special thanks are due to Nicolas Baumard, Pascal Boyer, Robert Brandenberger, Jean Bricmont, Coralie Chevalier, Jerry Fodor, Randy Gallistel, Rochel Gelman, Jacob Feldman, Julie Franck, Len Hamilton, Pernille Hemmer, Kathy Hirsh-Pasek, Celeste Kidd, Barbara Landau, Dan Ogilvie, John McGann, Melchi

Michel, Steve Piantadosi, Manish Singh, Paul Smolensky, Dan Swingley, Deena Weisberg, and the late Marc Jeannerod. I am also indebted to Victor J. Stenger, who read a draft of the manuscript and provided many insightful comments and helpful suggestions. Vic was one of my intellectual heroes; he was a brilliant physicist, a fellow nonbeliever, and I am honored that he agreed to write a foreword to *The Soul Fallacy*. Sadly, Vic passed away at the end of the summer. He was a great mind, an amazingly prolific writer, a passionate advocate of science and reason, and an extraordinarily kind and generous person. He will be sorely missed.

I also benefited immensely from the teaching and research environments at the Center for Cognitive Science and the Psychology Department at Rutgers University. The freedom I have been able to enjoy there has allowed me to teach classes on some of the topics discussed in the book. The approach I adopted in writing *The Soul Fallacy* was very much influenced by the thoughts, reactions, comments, and questions that I received from the students in those classes. Their feedback was an invaluable part of my own learning and writing experience. Discovering which ideas and conclusions resonate with nonspecialists and which ones prove to be more challenging helped me to adjust the level of the discussion and better understand the needs of the general public. I am indebted to all those students, and I am particular grateful to Hayley Fitzgerald and Sonia Skooglund, who helped me collect some of the data presented in the book, and to Melanie Kelliher, Cassandra McLean, Nisa Qais, Kathy Pindych, and Paige Seifert, for their assistance with various aspects of this project.

Finally, I would like to express my deepest appreciation and gratitude to my wife, Dianna, who provides a vital anchor for my heart and without whose love and support this project would have remained a distant dream. I dedicate this book to her.

Chapter 1

LIFTING THE VEIL

Every atom in your body came from a star that exploded. And, the atoms in your left hand probably came from a different star than your right hand. It really is the most poetic thing I know about physics: You are all stardust.
—Lawrence Krauss, *A Universe from Nothing*, 2012

One July night in a small English village, sometime near the end of the twentieth century, Harry stood by his friend Rodrick as the radio engineer calmly explained his plan to strike at the creator of the universe. Rodrick had decided that he wanted to kill God, and he thought he knew how. This desire was motivated in part by his conviction that the universe should exist on its own, but mostly it was fueled by Rodrick's deep contempt for the unfairness of existence for which he held God responsible. He explained to Harry that even though God was not material, He must possess at least some material characteristics, for otherwise He would not have been able to create the physical universe. When prompted to explain how he might be able to reach God, Rodrick remarked that the information had been available to us for a long time: "And God said, Let there be light: and there was light," (Genesis 1:3).

The machine that Rodrick built to carry out his plan was an elaborate framework of lasers, mirrors, and prisms, all precisely arranged and calibrated, sitting on the workbench in his home laboratory. He reasoned that it should be possible to generate a self-sustaining pattern of light that would reinforce itself indefinitely, transcending space and time to reach the Creator, striking God with a deadly bolt of energy. The two men adjusted their goggles and Rodrick flipped on the switch. Through the dark lenses, they could make out the pattern of light in front of them as the beams followed their geometric paths. Gradually, the light intensified, and the brightness started to expand, swallowing the mirrors, the workbench, and the entire room. An instant later, the light was gone. "That's it," announced Rodrick dryly. "God is dead."

Harry looked around, and everything seemed perfectly normal. "Nonsense!" he snarled. Rodrick then removed his goggles to inspect the room, and it was at that moment that the truth was revealed to Harry. He saw his friend's empty eyes. . . . Rodrick had indeed killed God, and in the process, he had destroyed every living creature's soul. Life went on, and the vast clockwork of the universe continued to tick according to mechanical laws, but all you had to do now was look into people's eyes to realize that they were all dead inside. There was no beauty, no meaning, no inner life. This is what God supplied when he was alive, after all, reflected Harry. And now it was all gone.

This is a summary of the short story called *The God Gun*, by science fiction author Barrington Bayley, which was written in the early 1970s.[1] Today, in spite of considerable advances in technology, most people would find Rodrick's quest futile and hopelessly simpleminded, to say nothing of its evil nature. But Bayley's story remains powerful because most of us share his intuition that human beings are more than mere collections of physical parts. There must be something else in addition to the atoms and cells that make up our bodies—an essence, a spirit, something precious and beautiful. In short, a soul. This intuition is deeply rooted in the human psyche and has been shared by people across cultures from antiquity to the present day. As Mark Baker and Stewart Goetz observe in their book *The Soul Hypothesis*, "Most people, at most times, in most places, at most ages have believed that human beings have some kind of soul."[2]

This intuition also plays a central role in most religious doctrines. Pope John Paul II famously articulated the idea in a message delivered to the Pontifical Academy of Sciences in October 1996, in which the Holy Father declared that the human body might originate from preexisting living matter, but the spiritual soul is a direct creation of God. Explaining the mind as a product of evolution, claimed the pope, was incompatible with the truth about man.[3] Belief in the soul is also very much alive in North American culture today, as the results of numerous polls demonstrate (we'll go over the details in the next chapter). In my own interviews of college students enrolled in upper-level undergraduate psychology classes like the ones I regularly teach at Rutgers University, I have found that a majority of students also believe that they have a soul. What's more, these intuitions are constantly reinforced by a wealth of books, TV shows, movies, and pronouncements made by writers and gurus of all stripes who purport to have found convincing evidence for the existence of the soul. Belief in the immortality of the soul was even featured as the cover story of the October 15, 2012, issue of the magazine *Newsweek*, with the title *Heaven Is Real: A Doctor's Experience of the Afterlife*.

In sharp contrast to popular opinion, the current scientific consensus rejects any notion of soul or spirit as separate from the activity of the brain. This is what Francis Crick, codiscoverer of the structure of DNA, called "The Astonishing Hypothesis."[4] In Crick's words, "You, your joys and your sorrows, your memories and your ambitions, your sense of personal identity and free will, are in fact no more than the behavior of a vast assembly of nerve cells and their associated molecules." Reflecting on what he calls the scientific image of persons, the philosopher Owen Flanagan stressed that we "need to demythologize persons by rooting out certain unfounded ideas from the perennial philosophy. Letting go of the belief in souls is a minimal requirement. In fact, desouling is the primary operation of the scientific image."[5] The weight of the scientific consensus is distributed over many disciplines and includes, as we would expect, the sciences of the mind (psychology, neuroscience, cognitive science). Harvard psychologist Joshua Greene summarizes the situation as follows:

> Most people are dualists. Intuitively, we think of ourselves not as physical devices, but as immaterial minds or souls housed in physical bodies. Most experimental psychologists and neuroscientists disagree, at least officially. The modern science of mind proceeds on the assumption that the mind is simply what the brain does. We don't talk much about this, however. We scientists take the mind's physical basis for granted. Among the general public, it's a touchy subject.[6]

Thus, according to Greene, science, like Rodrick's God-gun, has killed the soul, but scientists are reluctant to announce the news. The soul may indeed be a grand illusion, but it is a useful and comforting one. Open Pandora's box and we may be the ones, like Harry, looking into other people's eyes and discovering that everything has lost its beauty and meaning.

The award-winning author Jared Diamond once remarked that science is responsible for dramatic changes to our smug self-image. Astronomy has taught us that our planet is not the navel of the universe. We learned from biology that we were not created by God but evolved alongside millions of other species. This book is about another seismic change in our self-image. Most people today believe that we have the bodies of beasts and the souls of angels. Science tells us otherwise. In the pages ahead, I will take you on a tour of history, philosophy, and science to show you that the soul, like geocentricism and creationism, is a figment of our imagination, and I will try to explain to you what gives rise to the illusion. Modern astronomy and the theory of evolution did not precipitate the end of the world. They are unmistakable signs of progress. Likewise, I will show you that in spite of repeated claims to the contrary, we lose nothing by letting go of our soul beliefs and—better—that we even have something to gain. It is this empowering conclusion that I want to leave you with as you reach the end of this book.

THE GHOST IN THE MACHINE

In a 1999 *Edge* debate featuring the biologist Richard Dawkins and the psychologist Steven Pinker, titled *Is Science Killing the Soul*, Dawkins pointed out that the word *soul* has different senses. One is the traditional idea that there is something incorporeal about us, that the body is spiritualized by a mysterious substance. In this view, the soul is the nonphysical principle that allows us to tell right from wrong, gives us our ability to reason and have feelings, makes us conscious, and gives us free will. Perhaps most important, the soul is the immortal part of ourselves that can survive the death of our physical body and is capable of happiness or suffering in the afterlife. This is the soul that this book is about. It is the soul that captures the imagination of a majority of our population. Here's what some of the students I interviewed wrote about it:

Soul to me is the internal self of an individual. It's separate from the physical part of the body and makes me what I am. It is what I refer to when
I am thinking or talking about myself. . . . I do believe that my soul will
survive the death of my body. I think soul is eternal and will still be there
long after my body has perished.

I believe my soul is the non-material being of myself. The part that is distinct from both my mind and my external body. I believe the soul to be
unchanging and eternal. . . . Because I think the soul is imperishable I also
believe that it will survive the death of my body.

I would define my soul as the spirit inside of me that is currently present
in a human form. The properties of the soul are that it contains all of our
emotions and feelings. I believe that when I die my soul will live on.

There are, of course, other senses of the word *soul*. One has to do with
emotional or intellectual intensity, as in "their performance lacked soul."
The word *soul* is also used metaphorically in a variety of expressions such as
soul mate, *soul food*, *soul music*, *soul searching*, or *lost soul*, to name just a few.
However, speaking of a performance that lacks soul, or of the poor souls that
perished when the *Titanic* went down, does not commit you to a particular
metaphysical view. Likewise, exclaiming "Oh my God!" upon realizing
that the value of your stock portfolio has plummeted does not make you a
religious zealot (if anything it makes you a materialist, albeit not one of the
kind that we will be concerned with here). For these reasons, I will have
nothing to say about these other senses of the word *soul* apart from pointing
out, as Dawkins did in the *Edge* debate, that they are terms that exist, but that
they are not the subject of this book.

The doctrine underlying the traditional notion of the soul is the view
known in philosophical jargon as substance dualism, sometimes also called
Cartesian dualism, after the seventeenth-century French philosopher René
Descartes. Descartes famously argued for the existence of two fundamentally different substances: the physical matter of bodies and the spiritual stuff
of souls. In Descartes's system, souls and bodies causally interact. Your soul
pushes your buttons, so to speak, and makes you do the things that you do.

Conversely, what happens to your body is felt, or experienced, in your soul. We will refine this picture of the soul, trace its historical roots (i.e., the evolution of the concept), and define the soul hypothesis in more detail in chapter 2. For the time being, suffice to say that my argument against the soul (understood as an immaterial, psychologically potent, and immortal aspect of ourselves) will also be an argument against substance dualism—the entirely different nature of mind and body necessary to allow for the existence of souls in the first place.

THE SOUL OF THE MATTER

So how can we decide whether souls exist? Is this even a question about which science has anything to say? To many people, the answer to my second question is a resounding "No." After all, science deals with phenomena that can be objectively observed and measured. The soul, by contrast, cannot be observed or measured because it is claimed to be immaterial. Therefore, soul beliefs belong to the realms of religion and metaphysics. This conclusion, however, as I will argue in more detail in the next chapter, is mistaken. The soul *is* a scientific hypothesis about the design and functioning of human beings (the stuff of biology, psychology, and neuroscience), and dualism makes claims about the detachability of mind and body and the existence of a substance capable of causal interaction with ordinary matter (the stuff of physics). As such, souls are fair game for scientific investigation, subject to the same criteria that apply to the evaluation of any other scientific idea (a line of reasoning developed more generally for other supernatural concepts by the physicist Victor J. Stenger in his book *God: The Failed Hypothesis*). After all, science can tell us what happened a fraction of a second after the big bang took place, some 13.8 billion years ago, when no one was around to make measurements or record anything. Is it so far-fetched that science would also have something to say about what we are made of and how we function?

Imagine an episode of *CSI: Miami* in which the investigators have a suspect and are beginning their forensic work. As they gather the evidence, they discover that the suspect has no serious alibi for the night of the crime. It also doesn't take them long to figure out what could have motivated their suspect to kill his victim. Using more sophisticated equipment, they uncover physical evidence that links the suspect to the victim and to the crime scene—blood stains on the suspect's clothes that match the victim's blood type, DNA evidence that positively identifies the blood as that of the victim, soil on the suspect's shoes whose chemical composition matches that of the soil at the crime scene. Our investigators are also able to get their hands on a recording of the crime scene taken by a surveillance camera at a critical time right before the crime, and using digital video-enhancing techniques, they manage to retrieve meaningful evidence. Aided by powerful face-recognition software, they are able to place the suspect at the crime scene at the right time.

When taken in isolation, few, if any, of the clues uncovered by our team of forensic experts are really incriminating. The soil on the suspect's shoes could have come from a walk he took at the scene of the crime the night before the victim was killed. If the victim and the suspect knew each other, perhaps they spent some time together before the murder took place, and the victim, who was subject to frequent nose bleeds, ended up accidentally soiling the suspect's jacket. Since the jacket was dark, the suspect didn't notice the blood stains until after the police apprehended him the following day. As for the other pieces of evidence, I am sure that you can easily concoct a plausible story as well. It is when taken together, however, that all these disconnected pieces of evidence acquire their collective power, and as they accumulate, we soon reach a point where it is no longer reasonable to conclude that our hypothetical suspect is innocent.

The forensic-investigation analogy is a good one because it also captures the story of how mainstream science has come to the conclusion that human beings most likely do not have souls. One of my goals in this book is to present all the relevant pieces of evidence—from psychology, biology, neuroscience, philosophy, and the physical sciences—to support the conclusion that, when considered collectively, they undermine the soul hypothesis to the point of oblivion. Notice that the conclusion, if we want to be intellectually honest, should not to be that *there is no soul*, but rather, that there are no good reasons to believe that we have souls, and that there are very good reasons to believe that we do not have souls. To anticipate my conclusions, I will show you that the soul has shrunk as scientific understanding progressed, that there is no objective evidence supporting the soul hypothesis, that there is no known formalism that describes the soul substance, that souls fly in the face of what we know about modern science, and that no explanatory gain comes from postulating the existence of souls. In sum, I will show you that the soul has exactly the set of properties that it should have if it didn't exist.

Just like other false ideas we entertained in the past had harmful consequences and stifled progress, so too, I will argue, do soul beliefs. In medieval Europe, during the Great Plague, people often had their lips sewn shut and their tongues cut off for fear that they would blaspheme and offend God. This was a perfectly rational practice, if brutally sadistic, based on a deeply flawed theory. Replace God's wrath by the germ theory of disease and the sadistic practice loses its raison d'être. In less dramatic fashion, but in the same conceptual vein, I will show you that our soul beliefs get in the way of a more humane society. Our dualist intuitions lead to beliefs that cloud important societal debates, such as abortion, stem-cell research, and the right to die with dignity. Our intuitive notion of justice, and therefore our entire criminal justice system, which is unusually harsh and biased in the United States, as we will discover, may also be premised on dualistic assumptions. These enormously important issues that we face as a society should be approached armed with the best knowledge we have, not with traditional ideas that have no scientific credibility.

Through the pages of this book, I will lead you on a journey through science and thought to arrive at the following conclusions:

- The traditional soul is as much a scientific hypothesis about our design and mode of functioning as it is a metaphysical or theological claim. Consequently, determining whether or not we have a soul is an objective endeavor that falls within the scope of science.
- In spite of many claims to the contrary, there is in fact no credible evidence supporting the existence of the soul.
- Modern science gives us every reason to believe that we do not have souls.

- Nothing gets lost, morally, spiritually, or aesthetically by giving up our soul beliefs. In fact, we even have something to gain.
- The scientific image of personhood, so feared and vilified in the United States, provides the basis for an empowering and practically beneficial alternative to the soul myth.

TONE AND TACTICS

On a general level, the case against the soul is similar to the argument against the luminiferous ether of the nineteenth century, an invisible substance with mysterious properties, which was believed to serve as the medium for the propagation of light. The ether was an idea that was once entertained by the most serious scientists, but as understanding progressed, the need for such a substance became superfluous, and the ether hypothesis was eventually abandoned. From an emotional standpoint, however, the unraveling of the ether and the demise of the soul are as different as night and day. Most people did not have an opinion, let alone feelings, about whether the ether was real. Nothing about their lives hinged on the existence of the ether, and sacred doctrines did not contain divine prescriptions regarding the ether and its metaphysical significance. When it comes to the soul, it is a completely different story. For many people, the existence of an immaterial soul forms part of an intimate set of convictions and provides the basis for a deeply meaningful worldview. In no small sense, for such people, belief in the soul is a matter of life and death (literally so for those who believe in an afterlife).

These considerations bring up an important issue that has been regularly discussed within scientific and skeptical circles: the issue of tone. In reflecting on this question, the late Carl Sagan, who has done so much for skepticism and the public understanding of science, observed that when skepticism is applied to issues of public concern, as in the present case, there is all too often a tendency to belittle, to condescend, and to disregard the fact that believers are human beings as well, with genuine beliefs and real feelings, people who, like skeptics and scientists, are also trying to understand the world and figure out what their place and purpose in it might be.

Echoing Sagan's concerns, the astronomer Phil Plait delivered an address at The Amazing Meeting (TAM) of July 2010 titled, "Don't Be a Dick" (a maxim related to Wheaton's Law, which provides guidelines on appropriate online game-playing behavior, but that was also intended to apply to life in general). The gist of Plait's remarks was that even the best ideas are useless unless they are communicated. And in the case of skepticism, the message communicated has the potential to make people uncomfortable and defensive, to say the least. Consequently, our attitude and the way we communicate those ideas takes on critical importance.

I must confess that I have been guilty of the bias described above, and I was unaware of it until a student pointed it out to me when she wrote the following:

> I came into this discussion excited for this new point-of-view and eager to learn, but I remember leaving the lecture hall on the verge of crying. I know that dualism isn't the best explanation for the world around us, and it's good to hear both sides, but the way he explained it felt like daggers were being thrown in my heart and my world was shattering. I wish he would've let us down gently, like saying "Santa may not be here physically, but he'll always be in our hearts" instead of just yanking off the beard on the mall Santa and yelling in front of all the little kids, "SANTA ISN'T REAL!"

This is beautifully put and painful to read, and I felt sincerely sorry for eliciting such feelings. Those remarks also provided an important reality check. Since then, I have become much more sensitive to the issue of tone, and I have made a conscious effort to bear this in mind whenever I discuss the issue of the soul publicly or write about it. Tone, therefore, is something I will be sensitive to in this book. In doing so, I am reminded of Spinoza's motto, a dictum named after the seventeenth-century Dutch philosopher Baruch Spinoza and expressed in these words: "I have made a ceaseless effort not to ridicule, not to bewail, not to scorn human actions, but to understand them." In this regard, I also wish to make it clear at the outset of our investigation that this book is not intended as another broad-brush critique of religion, any more than a condemnation of drunk driving should be construed as a general diatribe against the use of motor vehicles. I am interested in the soul not because it is a religious concept and I have a bone to pick with religion but because it represents a fundamental aspect of human psychology.

Truth be told, there is a small group of soul advocates whose ideas I will criticize quite overtly in the pages ahead. These are the authors of popular books claiming to show that science supports the existence of the soul. I call them the New Dualists. When I discuss their ideas, the tone will be more pointed, if only for rhetorical purposes, but the criticism will always be directed at the ideas themselves rather than at the individuals who proposed these ideas. Besides, the New Dualists are all seasoned writers, and so unlike regular folks, they are used to having their ideas critiqued. This is just part of the game and it comes with the territory. Needless to say, the same rules also apply to my own ideas. With only one exception, I do not personally know the New Dualists, but I am sure that they are a great bunch, and I would be happy to share a stage with them if the opportunity presented itself.

Finally, I am also aware of the fact that even if I manage to find the right tone, the ideas that I will discuss in this book, and especially the conclusions that I will reach, might be offensive and sacrilegious to some. Here lies the dilemma that one finds at the heart of the scientific enterprise. On the one hand, the advancement of knowledge and understanding is a mission of critical importance in any society, and consequently, it is an endeavor that should be undertaken with earnest conviction and zeal. On the other hand, science has the singular property of revealing to us nature's ways without the kind of sugarcoating that might sometimes be helpful. Reality, for better or worse, happens to be the way it is and not the way we would like it to be. Inevitably, certain conclusions are bound to rub us the wrong way, which is the price we need to pay for looking behind nature's curtain to take a peek at its true face.

Related to the issue of tone, when writing on a sensitive topic, is the issue of tactics. Philosopher Owen Flanagan describes three such tactics, which I paraphrase here.[7]

(1) You may say: "You are really naive to believe X; we'll have to educate you so you can think straight and let go of all that silly nonsense."

(2) You may say: "There are good reasons to believe that X is not true, but we are confident that Y is true, and Y is close enough to X that you'll eventually get used to it. As you can see, everything will be alright, and the world won't come to an end."

(3) Or you may adopt the following strategy: People usually speak of X meaning X, but when you, the skeptic, speak of X, you really mean Y, hoping that your intended meaning will win the day, so that others will eventually come to mean Y when they talk about X.

I feel that (1) would simply be the wrong approach, for all the reasons I mentioned when I discussed the issue of tone. I also find (3) somewhat disingenuous. So (2) then will be my strategy of choice.

WHEN THE SPIRIT MOVES YOU

What would possess someone to publicly blurt out, like the child in Andersen's famous tale, that the emperor has no clothes, and worse, that he has no soul either? One of my favorite answers comes from one of my colleagues who once said, when asked a similar question: "I am paid to find out the truth and announce it!" (To be fair, this remark was probably made tongue-in-cheek, and besides, not all truths are born equal.) For those of us who are involved in the business of teaching psychology, neuroscience, or cognitive science, the soul certainly represents a perfect illustration of the proverbial elephant in the room.

We cognitive scientists routinely talk about the physical basis of mind and use phrases such as "the mind is what the brain does." Much less often do we publicly discuss what the physical basis of mind entails for the traditional notion of personhood. This is no doubt in large part because, as Joshua Greene pointed out, the question of the soul is a touchy issue. But just because an issue is touchy doesn't mean that we shouldn't talk about it. In fact, if we really are in the business of education, we should talk about such issues *precisely* because they are touchy and therefore rarely discussed publicly. After all, clergymen, movie directors, and politicians openly talk about the soul, so why shouldn't scientists?

It should go without saying (but it goes even better if we say it, as one of my high school teachers liked to remind us) that the goal of such discussion isn't to bully people who happen to believe in the soul into changing their beliefs. Rather, the objective is to create a free marketplace of ideas, where all points of views can be discussed without fear of censorship or discrimination, and to let people decide for themselves which set of ideas they find the most compelling. If teachers, educators, scientists, and writers were discouraged from discussing touchy, unfashionable, or controversial topics on the grounds that they are, well, touchy, unfashionable, or controversial, then education, like Harry and Rodrick's world, would lose much of its value and meaning.

Ironically, fairness and the recognition of different points of view is precisely what is often called for by proponents of certain "controversial" ideas in America today. Take for example the perennial "debate" over creationism and evolution that has been raging in the United States for many decades (much to the astonishment of our European friends). One of the arguments often made by proponents of intelligent design (the latest brand of creationism) is that we should be fair and teach students both sides of the "controversy." "Teach the controversy and let the students decide for themselves!" we often hear (sometimes from people as prominent as the president of the United States, in the case of George W. Bush[8]). Teaching the "controversy" in the evolution vs. intelligent design "debate" would be an excellent idea indeed if there actually was a meaningful controversy in the first place. To be sure, there is a huge manufactured, and largely North American, public controversy, but it has no analogue in the scientific world (hence the scare quotes when I used the words *controversy* and *debate*).

In the case of the soul, if there is a public controversy over its existence at all, it has been a pretty quiet one, at least compared to the battles raging over evolution. Nevertheless, while a substantial majority of the American public believes in the soul and its survival after death, mainstream science has abandoned this traditional idea. So here we have two worldviews that could not be more different from one another, and if we really care about being fair and ensuring that different ideas get their share of airtime, I say it's time to give scientists the microphone. As the psychologist Paul Bloom put it: "Such issues are too important to leave entirely in the hands of lawyers, politicians, and theologians."[9]

This book is the rejoinder to the growing number of popular books that have surfaced in recent years, trying to make the case for the soul on scientific grounds. Examples include *Life after Death: The Evidence*, by conservative writer and Christian apologist Dinesh D'Souza; *Life after Death: The Burden of Proof*, by New Age author Deepak Chopra; *The Soul Hypothesis: Investigations into the Existence of the Soul*, by linguist Mark Baker and philosopher Stewart Goetz; *The Spiritual Brain: A Neuroscientist's Case for the Existence of the Soul*, by neuroscientist Mario Beauregard and journalist Denyse O'Leary; and *Proof of Heaven*, by neurosurgeon Eben Alexander. Here's a revealing passage from D'Souza's book:

> To reclaim the hijacked territory, Christians must take a fresh look at reason and science. When they do, they will see that it stunningly confirms the beliefs that they held in the first place. What was presumed on the basis of faith is now corroborated on the basis of evidence, and this is especially true of the issue of life after death. Remarkably, it is reason and science that supply new and persuasive evidence for the afterlife—evidence that wasn't there before.[10]

So, according to D'Souza, science itself provides persuasive evidence for the immortality of the soul. If so, one might wonder why mainstream scientists themselves are not convinced by the kind of evidence that D'Souza claims exists. In fact, the scientific consensus goes precisely in the opposite direction: away from the soul and the afterlife—as we will discover in chapter 2. And it's not that D'Souza's fellow Christians failed to notice these developments. Consider, for example, the following passage from the back cover of a 2004 book titled *What about the Soul? Neuroscience and Christian Anthropology*, edited by the theologian Joel Green:

> Everyone knows about the rocky relationship between science and theology brought about by the revolutionary proposals of Copernicus and Darwin. Fewer people know about an equally revolutionary scientific innovation that is currently under way among neurobiologists. This revo-

lution in brain research has completely rewritten our understanding of who we are. It poses fundamental challenges to traditional Christian theology. According to the scientific worldview that now dominates, it is no longer necessary to speak of a soul or spirit as distinct from the functions of the brain.[11]

Contrary to what D'Souza and others have claimed, I passionately disagree (perhaps I should say that I rationally disagree) with the conclusion that science supports the notion of an immortal soul. As I will argue in the pages ahead, the current scientific consensus isn't simply a fad, nor is it fueled by antireligious sentiment (as Baker and Goetz suggest in their book). Instead, scientists have abandoned the soul because reason and evidence—the tools of their trade—compelled them to do so.

THE TRUTH IS OUT THERE

Although the world rarely comes neatly prepackaged into clearly delineated categories, and contains many shades of gray, I have come to see that people fall into roughly three categories with respect to the question of the soul. At one end of the spectrum are my colleagues and fellow scientists, for whom the conclusions I will reach in this book will not be particularly surprising. At the other end of the spectrum, we find people who are intimately convinced that human beings have souls and who will not consider, even in principle, that this may not be true. I've often heard them politely tell me that scientists can present all the evidence they want, no amount would ever convince them to change their minds. Just like Agent Mulder in *The X-Files*, they *want* to believe. Period.

There is also a third category. These are people who sit on the fence regarding the existence of the soul, although they may be leaning one way or the other. At the heart of their dilemma lies the massive asymmetry that we find in a country like the United States—for every bit of information that is released from the ivory tower and reaches the general public on the topic of the soul, there are ten gigabytes of countervailing information pouring out from all corners of popular culture. So those undecided souls (here's an example of usage that does not carry any metaphysical implications) end up being immersed in the traditional view, but they are only vaguely familiar with the details of the scientific view. I have met many of those people myself and come to realize that they, unlike the people in the previous category, are in principle willing to change their minds if someone takes the time to carefully explain to them why it is that mainstream scientists no longer believe in the soul. Sometimes, a little rational push is all it takes to awaken the Agent Scully within.

This book is equally suited to readers who fervently believe in the soul and will not change their minds, but who are nevertheless curious and would like to try to understand why there are people who do not believe in the soul narrative. As for the choir of colleagues and fellow scientists to whom I would be preaching, I am reminded of my own experience with ideas that I already accept. For example, I do not need to read any more books to convince myself of the validity of evolution, but I still read books on this topic because I find the details fascinating. I am also a big sucker for analogies that help convey complicated or important ideas (or both) in simple and compelling ways (the philosopher Daniel Dennett calls these *intuition pumps*, and I will use several of them throughout the book). So if other academics are like me in this regard, and I suspect that many are, I am sure there are many aspects of this book that they will enjoy too, even if the denouement is a forgone conclusion.

STAIRCASE TO THE STARS

Now that we are clear on the goal, the approach, and the motivations, let me tell you how I am going to proceed. The next two chapters set the stage for our investigation and give us the tools we need to embark on our soul-searching journey. In chapter 2, we will explore the soul hypothesis. We'll begin by taking a historical perspective on the question of the soul and retrace the evolution of the dualistic doctrine going all the way back to ancient Egypt. Along the way, we will become acquainted with what thinkers like Plato, Aristotle, Democritus, and Descartes had to say about the soul.

This brief voyage through time will reveal that our modern conception of a unitary soul began its life as a plurality of souls, each responsible for different biological and psychological functions, and that as scientific understanding progressed, these souls began to melt away, like snow in the sun. But the soul hasn't completely disappeared—far from it. Numerous polls and surveys indicate that most people in the United States today, and many more around the world, believe in the existence of an immaterial, psychologically potent, and immortal soul. To buttress this conclusion, I will share with you the results of my own investigations, showing that even students taking upper-level undergraduate courses in psychology believe in the existence of a soul that closely resembles the Cartesian one—precisely the kind of soul that modern science has abandoned. Armed with a clear definition of the soul, we will formulate our two competing hypotheses, the dualistic and materialistic conceptions of personhood. I will then show you that the soul hypothesis, contrary to what many people believe, represents a bona fide scientific claim. Before beginning the heavy detective work, we will need to familiarize ourselves with our thinking tools and learn to recognize the many ways in which our minds can err. This will be our main task in chapter 3. As physicist and Nobel laureate Richard Feynman wrote, "The first principle is that you must not fool yourself, and you are the easiest person to fool."[12] Our task, then, as followers of Feynman's first principle, will be to make sure that we can avoid fooling ourselves as we tackle the question of the soul.

Chapters 4 through 6 provide an in-depth evaluation of the soul hypothesis. Chapter 4 presents the first line of reasoning that has led modern science away from the soul. We will discover that in spite of well-publicized claims to the contrary, there is in fact no credible evidence for the existence of the soul. I will show you that the authors who assert otherwise, a group of people I call the New Dualists, are simply not playing by the rules of the scientific game. Instead of trying to convince the scientific community of the validity of their conclusions, the New Dualists pitch their extraordinary claims directly to the general public, banking (sometimes quite profitably) on the fact that nonspecialists will not be able to tell the difference between good science and bad science. But after reading chapter 3, you will be able to tell the difference and recognize when someone is trying to pull the wool over your eyes. To debunk the New Dualists, we will focus on four families of claims typically invoked to make the case for the soul. We will examine the alleged powers of introspection (the self-examination of one's conscious thoughts and impressions), the fascinating world of near-death experiences, the limits of science, and the interpretation of established scientific conclusions. In each case, I will show you why the evidence is not credible and explain why mainstream science should not—and does not—take it seriously.

In chapter 5, we will consider the second line of reasoning that has led mainstream scientists to abandon the soul. I will show you that the notion of an immaterial substance that can causally interact with our body to give us our psychological powers flies in the face of what we know about modern science. In the process, we will discover that the dualistic doctrine rests on conceptual quicksand, and that the very notion of an immaterial or non-physical soul substance is either nonsensical or trivially reduces to physical factors. Our overarching conclusion will be that the soul has exactly the set of properties that it should have if it didn't exist.

Chapter 6 drives the final nail in the soul's coffin. Having shown you that there is no credible evidence for the dualistic hypothesis, and that the soul doctrine flies in the face of what we know about modern science, I will present the evidence supporting the materialistic alternative. To survey the landscape, I will take you on a tour of cognitive neuroscience and review classic results, evidence from unusual syndromes, as well as cutting-edge research relying on the latest developments in brain-imaging techniques.

The last three chapters focus on the implications of the demise of the soul. Chapter 7 explores the roots of the soul illusion, the conception of mind that emerges in a soul-free world, and the vexing problem of consciousness. To unravel the soul illusion, we will immerse ourselves in the world of cognitive development and learn that human infants distinguish between body and soul very early in their development. As psychologist Paul Bloom argued in his book *Descartes' Baby*, we are innate dualists. I will also show you that explicit dualistic beliefs emerge early in development and can already be found in preschool children. We will discover that there are also good reasons, based on everyday experience, to erroneously conclude that our conscious mind is separate from our body. Finally, I will show you that supernatural concepts like the immortal soul are minimally counterintuitive concepts in the sense that they are minor variations on the natural concepts that we routinely entertain. This will help us understand some of the characteristics of soul beliefs as well as their cultural prevalence.

Next, we will turn our attention to the modern conception of mind. If the mind isn't a separate substance, then what is it? And if our mental lives emerge from physical activity in our brains, as we will discover in chapter 6, then why should we talk about the mind at all? To answer these questions, I will show you that talking about the mind is simply a way of talking about the brain at a certain level of abstraction. We will then come face-to-face with one of the deepest mysteries in the study of mind—the problem of consciousness. While consciousness continues to baffle scientists and philosophers, we will see that it provides no basis to rescue the soul doctrine.

In chapter 8, we will confront Harry and Rodrick's fears and conclude that they are ill founded. Nothing gets lost if we let go of dualism, because the soul was only a potential explanation for certain facts about the flexible behavior and the complex mental lives of human beings. Explanations change, but facts are here to stay. We will then turn to the implications of the demise of the soul for our conception of free will and moral responsibility. The problem we face, critics of materialism argue, is that if we remove the soul from the human equation, and thus dispense with our intuitive notion of free will, we lose any meaningful notion of moral responsibility. I will show you why this line of reasoning is mistaken and explain how we can preserve a meaningful conception of moral responsibility in a soul-free world. We will then turn to the question of immortality and conclude that the soul narrative is only one of several ways to cope with our own mortality. If so, we lose nothing by letting go of the soul.

In the final chapter of the book, we will turn the tables on the soul narrative and discover that dualistic beliefs actually stand in the way of progress and a more humane society. Because they inevitably distort our perception of reality, soul beliefs corrupt our thinking on matters that are deeply important to all of us. I will show you how soul beliefs subvert the intuitions that underpin our notion of moral responsibility, poison the debate over abortion, and muddle the issue of whether people should have the right to die with dignity. Having fully exposed the soul fallacy, I will then give you reasons to embrace the materialistic conception of personhood and offer you an empowering alternative to the soul myth. Our conclusion will be that truth and happiness are indeed compatible.

Returning to Rodrick, the dark inventor's quest might have been misguided, but he was nevertheless onto something important: the intuition that we owe our existence to death on a cosmic scale. In the beginning there was light, in the form of giant balls of gas, programmed to burn, and eventually self-destruct, by the very laws of nature that gave birth to them. The burning and dying of massive stars, through a process called stellar nucleosynthesis (which would have been every alchemist's dream), led to the creation of elements heavier than hydrogen (carbon, oxygen, nitrogen, and iron, to name just a few) of which we, and the world around us are composed. No prophet or sacred man could ever have dreamt of a story like this one because it is far too daring and improbable for anyone to have imagined. As the physicist Lawrence Krauss said, we are all made of stardust. To that, I would add that we are thinking, feeling, and loving stardust. It is this beautiful conclusion and its implications for the notion of personhood that I want to share with you in this book.

Chapter 2

THE SPIRIT OF THE AGE

Most people, at most times, in most places, at most ages have believed
that human beings have some kind of soul.
—**Mark Baker and Stewart Goetz,** *The Soul Hypothesis,* 2011

Edwin Smith, an Egyptologist from Connecticut, was born in 1822, the year Jean-François Champollion published the first translation of the Rosetta Stone hieroglyphs. In 1862, Smith purchased a manuscript roll in Luxor from a man named Mustapha Aga. Some of the outer portions of the manuscript were missing, but Smith was able to acquire the remaining fragments a couple of months later. The documents remained in Smith's possession until his death in 1906 when his daughter decided to donate them to the New York Historical Society. Two decades later, in a moment worthy of *Antiques Roadshow*, James Breasted, then director of the University of Chicago's Oriental Institute, deciphered the Smith Papyrus and revealed its secrets. The document predates the birth of Christ by seventeen centuries, although it is believed to be a copy of an even older text, and it offers an unparalleled glimpse of ancient Egyptian medicine.

Part of the historical value of the Edwin Smith papyrus comes from its neatly organized description of forty-eight case studies involving traumatic injuries to the head and neck. What makes the document truly remarkable, however, is that unlike other known Egyptian medical texts like the Ebers Papyrus or the London Medical Papyrus, the Smith Papyrus is not based on magic—none of the usual spells, potions, godly elixirs, or other divine incantations.[1] Instead, the document presents an uncharacteristically rational and thoroughly modern approach to Egyptian medicine, bordering on the sterile. Thanks to the wonders of the Internet, anyone can now peruse a translated version of the document and read about the fascinating cases it describes. In case 8, the Egyptian surgeon tells us about a man with a smashed-in skull, and comments on the poor fellow's shuffling walk. Case 22 mentions a man with a smashed-in temple who was rendered speechless by his injury. In case 31, we learn of a man with a dislocated vertebra in his neck who can no longer feel his arms and legs. Did the ancient Egyptians understand the implications of their observations for the relationship between mind and body?[2] Could the origins of Francis Crick's astonishing hypothesis, his scientific search for the soul, be traced back to ancient Egypt?

Anachronistic musings aside, it is doubtful that the Egyptians would have had much sympathy for Crick's hypothesis. Like most of us today, the people who gave us the great pyramids also believed in the soul and the afterlife, attesting to the timelessness of these ideas. For the Egyptians, the soul was composed of five parts[3]: the *ib*, the *ba*, the *ka*, the *sheut*, and the *ren*. The *ib*, or heart, was responsible for thought, will, and emotion. Upon death, Anubis, the jackal-headed god of the underworld, would examine the heart and weigh it with a feather to decide its fate in the afterlife. The *sheut*, a person's shadow, contained aspects of the person to which it belonged. Pharaohs even carried a box in which their *sheut* could be stored. The *ka* was a kind of vital essence believed to leave the body at the moment of death. The *ba*, which was also believed to survive the death of the body, represented someone's unique personality. The *ren* was someone's name, and it was believed to continue living as long as it was spoken.

The sophisticated medical knowledge of the Egyptians and their belief in a multifaceted soul illustrate important aspects of the age-old quest to unravel the mysteries of human nature. They also offer tantalizing clues to the origins of the tension between the two rival conceptions of personhood that we will explore throughout this book—dualism and materialism. Before we start our detective work, we will need to familiarize ourselves with these ideas, their origins, and their evolution. This will allow us to better frame the question that lies at the heart of our soul-searching journey. The late historian Howard Zinn insisted that understanding history is important. Those who don't know history, Zinn warned us, might as well have been born yesterday. To better understand the nature of soul beliefs, we will need to explore their historical roots. This will be our first task. History teaches us that soul beliefs are timeless, close to universal, and that they have been associated with the phenomena of life, mind, and death. Along the way, we will discover that the soul began its life as a plurality of entities that have undergone important transformations in the course of history to give rise to the kind of soul that most people are familiar with today.

Our second task will be to focus on the present. The history of soul beliefs is certainly fascinating, but there would be little point in writing a book like this one if people today no longer believed in the soul. It would be like trying to convince a group of meteorologists that Zeus doesn't cause thunder and lightning. In our modern world, the soul is far from an antiquated relic from the past. Today, the soul is alive and well, and it still matters to billions of people around the globe. To be sure, soul beliefs are part and parcel of almost every religious doctrine. To Christians and Muslims, few questions are more important than the fate of the soul. Will the deceased enjoy the eternal rewards of paradise or be doomed to perpetual torment in hell? The Jewish tradition is less focused on the afterlife, but the terms *nephesh* (living person) and *ruach* (wind) have been used to describe the soul. The soul is known as *atman* in Hinduism and Jainism, and Chinese Taoism divides it into two main parts, the *hun* (cloud-soul) and the *po* (white soul). Even animistic and tribal religious systems have soul-like entities such as spirits and ancestors.

The close family resemblance between soul claims and religious beliefs can be misleading however. Most people who believe in the soul today take it for granted that their dualistic doctrine is safely insulated from the corrosive reach of materialistic science because the soul is, by definition, *immaterial*. Consequently, soul claims are often presented as *religious* claims, propositions about which science has nothing to say. But this conclusion is mistaken. The soul *is* a scientific claim, in much the same way that statements about the age of the earth or the origin of species are scientific hypotheses. This means that the existence of the soul is a matter that can be investigated objectively. Either we have a soul, or we don't. And the best way to find out is to rely on the same method we use to decide whether a suspect is guilty or innocent (to return to my earlier analogy). In both cases, our judgment should be based on reason and evidence.

The scientific nature of the soul hypothesis also sheds light on a peculiar aspect of North American culture. Among the general public, dualism reigns supreme and belief in the soul represents the majority opinion. Push open the gates of the ivory tower, and you will find that most mainstream scientists have abandoned the soul. This is not a good sign. When science decides that an outdated worldview needs to be replaced, the writing is on the wall, and resurrection is rarely on the horizon. Think about the claim that the earth is at the center of the universe or that human beings were created separately from the rest of the animals. For modern science, the soul fallacy has been exposed, and the revolution has already taken place. All the same, the shock waves of this seismic change to our self-image have not yet reached the shores of public opinion. Before we start writing the soul's obituary, let us take a moment to commemorate the long and prosperous life it enjoyed.

EARLY SOUL BELIEFS

According to Arthur C. Clarke's third law, any sufficiently advanced technology is indistinguishable from magic.[4] For our prescientific ancestors, almost everything must have looked like magic. What distinguishes a rabbit— a furry lump of matter that breeds, runs, and jumps—from a rock—a much more boring hunk of matter that just sits there? In other words, what makes something alive? Where do our emotions and our capacity for choice come from? Can the matter in our bodies really *feel* and *think*? How come dreams feel so real to us, and why are they often so bizarre? An early answer to these mystifying questions was that different types of *souls* are responsible for life, thoughts, emotions, and dreams.

In his book *The Early Greek Concept of the Soul*, Jan Bremmer chronicles the evolution of our modern conception of the soul and shows that early soul beliefs represent our attempt to make sense of biological and psychological phenomena. Before there was one soul, *our* soul, *the* soul, there was a plurality of souls. There were *free souls* and *body souls*. Free souls could leave our bodies and transport us into the worlds of our dreams. These vagabond souls may be at the origin of *surviving souls*, the vehicles of our journey into the afterlife. Body souls, on the other hand, were more securely attached to us.

Those were *life souls*, which, as their name suggests, gave us the spark of life, and *ego souls*, to which we owe our sentience and rational powers.

Studies by Swedish religious historian Ernst Arbman (1891–1959) reveal that beliefs about free souls and body souls were widespread among native North Americans, the peoples of Northern Asia, Northern Europe, and India. English anthropologist Sir Edward Burnett Tylor (1832–1917), a pioneering figure in his field, concluded that soul beliefs can be found in virtually all early human cultures. More recently, George Peter Murdock (1897–1985), an American anthropologist known for his work on Old World populations, proposed a list of cultural universals that includes soul beliefs. In his book, Bremmer observes that the plurality of souls found in early cultures gradually led to the unitary notion underlying our modern conception of the soul.

This suggests that the soul's evolutionary history may be akin to that of a deflating balloon. As more aspects of nature yield to scientific understanding, the number of souls diminishes. With the passage of time and the march of progress, souls lose more and more of their turf, and remain alive only by virtue of our own ignorance.[5] Physicist Jerome Elbert points out that "the ideas for the different components of the soul arise from ignorance of how life, emotions, mental characteristics, consciousness, dreams, and trances arise in nature. In a technically advanced modern society, many of the items in the list have scientific explanations. In the long run, only those things that are not understood by science are apt to retain their supernatural explanations."[6] Our early ignorance of natural phenomena, combined with an intuitive commitment to a certain type of explanation, may be what gave rise to soul beliefs in the first place. We will explore these ideas in more detail in chapter 7.

As the work of historians and anthropologists reveals, soul beliefs were extremely common among the people of ancient cultures. Before they were molded into their modern form, however, conceptions of the soul had to pass through the sieves of history's greatest minds. While most people were content to just believe in souls, the great philosophers began to reflect on them in a more systematic fashion, profoundly altering the course of their evolution. Two of the most influential early thinkers on the soul were Plato and Aristotle. Plato believed in an immortal soul capable of surviving the death of the body. He wrote about it in *Phaedo* and *The Republic*, two of his many influential dialogues. Like most thinkers that followed in his footsteps, Plato believed that the soul is the principle that gives life to the body. For the great philosopher, the soul was a life soul, and it became a surviving soul after a person's death. Plato's doctrine maintained that the soul originates in the realm of the dead, temporarily passing through the world of the living, before it returned to the underworld in an endless cycle of reincarnation. Plato was one of the early dualists for whom the soul was a separate substance, distinct from the body, and capable of existing on its own.[7]

Plato also theorized about the structure of the soul and proposed that it is composed of three parts: the appetitive, the spirited, and the rational.[8] The appetitive soul was the lower of the three, the one responsible for our basic desires such as hunger, thirst, and sexual drive. On top of the hierarchy sat the rational soul, the principle that gives us our quintessentially human capacity for thought and reason. In the middle we find the spirited soul playing the role of a mediator between the appetitive and the rational soul. Plato even located the three parts of the soul in different areas of the body. The rational soul had its place in the head, the spirited in the heart, and the appetitive in the belly. In *Phaedrus*, Plato used the allegory of a chariot driven by two winged horses to describe the soul. The rational soul, the brave charioteer, had to muster all his might and wisdom to try to guide the two impetuous horses representing the appetitive and spirited souls.

Aristotle, Plato's brilliant student, also wrote about the soul in *De Anima*, his treatise on the nature of living things. Like Plato, Aristotle believed that the soul gives life to the body. All living things, Aristotle proposed, must have a soul. At the bottom of the hierarchy, we find the soul of trees and plants, the *nutritive* soul. This soul was responsible for the functions of nourishment, growth, and decay. One step up was the *sensitive* soul which gave nonhuman animals the senses of sight, smell, touch, taste, and hearing. And at the top of the hierarchy was the soul that distinguishes human beings from all other living creatures, the *rational* soul. So far, we can see the influence of Plato's ideas on Aristotle's thinking. But the two great men disagreed on the nature of the relationship between the body and the soul. As we saw, Plato was a dualist. He believed that the body and the soul are not cut from the same cloth and that one can exist without the other. Aristotle didn't.[9] For him, the soul was not a separate entity but rather a kind of form, or vital principle, which gave the inert body its properties.

Greece was the birthplace of many ideas, including a worldview called materialism, the soul's nemesis. Materialism is a monistic doctrine that postulates the existence of only one kind of substance, matter, to account for all natural phenomena, including life and mind. Democritus (460–370 BCE) and his teacher Leucippus believed that the world of matter is composed of tiny, indivisible particles they called atoms. According to the early materialists, the life soul of Plato and Aristotle was composed of one particular kind of atoms, fire atoms, perhaps due to the association of life with heat. Spherical fire atoms were restless little devils, and this could explain how the soul was able to cause the body to move.

The early materialists did not need to postulate a second substance, like Plato, nor did they rely on teleological (goal-oriented) or purposeful properties of matter like Aristotle's forms or vital essences. Everything, including life and mind, boiled down to the behavior of tiny atoms in motion. The early atomists, unlike the Egyptians, would have been very sympathetic to Francis Crick's astonishing hypothesis. Building on Democritus, Epicurus (341–270 BCE) and his followers proposed a thoroughly materialistic approach to knowledge that was a direct challenge to Platonism and an early attack on supernatural beliefs involving divine intervention. With the rise of the Roman Empire a few centuries later, and the rise of Neo-Platonic ideas, Epicureanism all but disappeared from the intellectual scene, until materialism made a comeback in the seventeenth century when French philosopher Pierre Gassendi (1592–1655), a contemporary of René Descartes, tried to reconcile Epicurean atomism with Christianity. Dualism and materialism have been at war with each other for a long time.

THE BIRTH OF THE MODERN SOUL

The Platonic and Aristotelian souls enjoyed a long period of prosperity, deeply influencing early Christian thinkers such as Augustine (354–430 CE) and Thomas Aquinas (1225–1274), until they underwent a major transformation at the dawn of the scientific revolution in the early seventeenth century. The individual responsible for pruning the soul down to its modern size was René Descartes (1596–1650), a Frenchman of modest stature but remarkable intellectual ability.[10] To some, it may be surprising to read that Descartes was involved in efforts to downsize the soul. After all, he is often regarded as the intellectual father of the soul, the man who gave us the most famous version of dualism. In reality, Descartes was one of the first thinkers to recognize that Aristotle's life souls, his nutritive and sensitive souls, were superfluous. Little did he know that his own soul would soon meet the same fate.

Descartes was a true polymath who wrote about physics, cosmology, biology, physiology, and psychology. As one of the fathers of the scientific revolution, he took a radical departure from the Aristotelian tradition, banning from our understanding of nature the quality-bearing forms and essences of Aristotle and his followers, the medieval scholastics. The early scientific revolution ushered in the mechanical philosophy, the idea that the world works like a giant machine. Descartes's mechanical thinking was heavily influenced by the automata at the Royal Gardens in Saint-Germain-en-Laye, a suburb of Paris. These hydraulically powered statues could perform all kinds of impressive tricks, and they served as the model that Descartes used to describe the functioning of the human body.

Descartes's mechanical man was an intricate assembly of tubes (blood vessels), pipes (nerves), and springs (muscles and tendons). The bodily machine was set in motion by the flow of "animals spirits" (a kind of fluid) stored in the brain and released by small valves that opened and closed when tiny wires inside the hollow nerves were pulled. Here's how Descartes describes the body in his *Treatise on Man*:

> I desire that you consider that all the functions that I have attributed to this machine, such as the digestion of food, the beating of the heart and

the arteries, the nourishment and growth of the bodily parts, respiration, waking and sleeping; the reception of light, sounds, odours, smells, heat . . . the impression of the ideas of them in the organ of common sense and the imagination, the retention or imprint of these ideas in the memory . . . I desire, I say, that you should consider that these functions follow in this machine simply from the disposition of the organs as wholly naturally as the movements of a clock or other automaton follow from the disposition of its counterweights and wheels.[11]

Viewing the body as a machine allowed Descartes to dispense with Aristotle's nutritive and sensitive souls. In the same passage, Descartes tells us why:

To explain these functions, then, it is not necessary to conceive of any vegetative or sensitive soul, or any other principle of movement or life, other than its blood and its spirits which are agitated by the heat of the fire that burns continuously in its heart, and which is of the same nature as those fires that occur in inanimate bodies.[12]

As a student of human nature, Descartes was also interested in the higher mental faculties of man, such as our capacity for rational thought and language. This is where the mechanical philosophy hit a major roadblock. How could the bodily machine with its fixed mechanical structure give rise to our flexible thinking and our creative use of language? (We will return to this question in chapter 4.) These properties seemed to fall beyond the explanatory reach of Descartes's material substance, *res extensa* (the extended substance). This led Descartes to postulate a second substance, *res cogitans* (the *thinking* substance) to account for the recalcitrant properties of mind. He was able to dispense with the nutritive and sensitive souls of his predecessors, but he could not do away with the rational soul.

Descartes's two fundamentally different substances form the basis of his enormously influential version of dualism. In his writings, Descartes tried to defend his view by showing that the body and the mind are as different as night and day. In one of the most famous thought experiments in the history of philosophy, Descartes imagined that an evil demon had fooled him into believing that the external world was real, whereas, in fact, it was nothing but a figment of his imagination. Be that as it may, there would still need to be a mind *thinking* that the world was real! This line of reasoning led Descartes to the idea that even if he could doubt everything else, he could not doubt his own existence as a *thinking* being. If it is possible to doubt that the body exists but *im*possible to doubt that the mind exists, then the body and the mind must be different things, Descartes concluded. Another piece of evidence Descartes offered to support his dualism of body and mind is the divisibility argument. Our conscious selves appear to us to be inherently unified, with no discernible parts. By contrast, the body is easily divisible into parts. We can metaphorically lose our head over someone, but we can also lose it quite literally, as Descartes's descendants demonstrated during the French revolution. If the body is divisible because it is material, then the indivisible mind must be immaterial.

Despite all his clever arguments to distinguish the mind from the body, Descartes still needed to find a way for his two substances to interact. After all, what good would the soul be if it couldn't push our buttons and make us do the things that we do? Descartes answered this question by specifying *where* in the body the soul exerts its causal powers. He proposed that the soul was able to interact with the body through a small structure in the brain called the pineal gland. On the more vexing question of *how* the ghostly soul could possibly affect the material body, Descartes remained rather vague. As we will discover in chapter 4, the interaction problem was Descartes's Achilles' heel. But let's not get ahead of ourselves. For the time being, the important conclusion is that with Descartes, the soul began to shrink. Plato's charioteer lost his winged horses and Aristotle's life souls evaporated, leaving the rational soul as the lone survivor of the mechanical philosophy.

THE PERIMETER OF IGNORANCE

In *The Soul Hypothesis*, Mark Baker and Stewart Goetz remind us that many of the greatest Western thinkers have been dualists, including Plato and Descartes, but also Leibnitz, Locke, and Kant, as well as brilliant scientists like Galileo and Newton.[13] In the twentieth century, philosophers such as Karl

Popper, and neuroscientists like Wilder Penfield (whom we will meet in chapter 6) and Sir John Eccles were also dualists. If such great minds believed in the soul, who are we to scoff at the idea? Someone could of course reply that other great minds, like Bertrand Russell and Albert Einstein, explicitly rejected the soul.[14] But this would be missing a more important truth obscured by Baker and Goetz's tallying tactic. Descartes was a visionary who recognized that the human body works like a machine. He may have been mistaken about the details, but his general outlook remains valid. When Descartes thought that he understood how something worked, even at an abstract level, he didn't need to postulate enigmatic notions like the life souls of his predecessors. But when he reached the limits of his own understanding, Descartes had no choice but to postulate his mysterious second substance.

Descartes's approach follows a more general pattern nicely captured by astrophysicist Neil deGrasse Tyson. In an essay titled *The Perimeter of Ignorance*, Tyson explains that scientists "invoke divinity [or magical substances] only when they reach the boundaries of their understanding. They appeal to a higher power only when staring into the ocean of their own ignorance. They call on God only from the lonely and precarious edge of incomprehension. Where they feel certain about their explanations, however, God gets hardly a mention."[15]

Consider the great Isaac Newton. Newton's law of gravity allows us to calculate the force of attraction between different bodies, like the earth and the moon. But what happens when additional bodies are introduced, like the other planets, all pulling and tugging on each other? Newton worried that this cosmic tug of war would destabilize the orbits of the planets in the solar system. His equations predict that the planets should have either fallen into the sun or flown off into space. But this is of course not what happened. Newton simply couldn't explain the stability of the solar system, and so, in his masterpiece, the *Principia*, he declared that God must directly intervene in the planets' business in order to maintain harmony in the solar system.

The stability problem was resolved about a century later by the great French mathematician and astronomer Pierre-Simon Laplace (1749–1827) in *Mécanique Céleste,* his five-volume treatise on celestial mechanics. Instead of relying on God, Laplace rolled up his sleeves and developed a new kind of mathematics called *perturbation theory* that allowed him to solve New-ton's multibody problem. Napoleon Bonaparte, who had a keen interest in science, acquired a copy of *Mécanique Céleste* but was surprised to dis-cover that the great mathematician never mentioned God in his work. When Napoleon asked why that was, Laplace is rumored to have replied "Sire, I had no need of that hypothesis."

Christiaan Huygens (1629–1695) was a prominent seventeenth-century thinker remembered for his studies of the rings of Saturn, his discovery of its moon Titan, his invention of the pendulum clock, and his wave theory of light. In 1696, Huygens published his own reflections on celestial mechanics in a book titled *The Celestial Worlds Discover'd*. Huygens's book is a repository of what was known about the solar system at the time. In the later chapters, Huygens marveled at the complexity of life and wondered about its origins. Since seventeenth-century biology was still in the dark about these ques-tions, Huygens cheerfully invoked the hand of God as he reflected on the intricate design seen throughout the biological world.

The diversity and complexity of life remained a mystery for another two hundred years or so, until Charles Darwin proposed his revolutionary ideas. Roughly a century later, Francis Crick and James Watson discovered the molecular structure of DNA, the genetic code of life. By the middle of the twentieth century, vitalism—the doctrine that living organisms are fundamentally different from inanimate matter because they contain a mys-terious nonphysical substance—had seen its last days. The physical basis of life, as Descartes predicted, rendered the life souls of Plato and Aristotle superfluous. As our ignorance of nature's ways diminishes, the soul shrinks. The last scientific frontier today is the human mind, and not surprisingly, this is where people are now looking for the soul.

THE SOUL TODAY

Descartes wrote about the soul more than three hundred years ago. Quite a few things have changed since the seventeenth century. What about soul beliefs? Let's begin by looking at a country that often prides itself as the most advanced and enlightened nation on earth, the United States. Who believes in the soul in today's America? Most people do, it turns out. In 2003, the Barna Research Group of Ventura, California found that 81 percent of Americans believe in an afterlife and that 76 percent believe in heaven.[16] Gallup polls taken between 1997 and 2004 reveal that belief in heaven among Americans ranges between 72 percent and 83 percent.[17] A 2008 Pew report on religious beliefs and practices found that 74 percent of Americans believe in an afterlife, with a majority of these people stating that they are absolutely certain of this belief.[18] According to a 2009 Harris poll, 71 percent of Americans believe in the survival of the soul after death.[19] Harris ran a follow up in 2013 and found that a solid majority of Americans, 64 percent, continue to believe in the immortality of the soul.[20]

Belief in the soul and the afterlife has been remarkably steady in the United States during the second half of the twentieth century. In 1947, on the heels of the Second World War, 68 percent of the population reportedly believed in the afterlife. In 2001, at the dawn of the twenty-first century, the number reached 76 percent, with very little change in the intervening decades. The prevalence of soul beliefs in the United States contrasts with attitudes reported in developed nations like France, Germany, the Netherlands, Belgium, and the Scandinavian countries. While the numbers in several of these countries were comparable to their North American counterpart in the middle of the twentieth century, they have since declined. Nevertheless, between 30 percent and 50 percent of the population in those countries remain committed to the afterlife.[21] In Africa and India, the numbers are even higher than in the United States. In his book *Immortality*, philosopher Steven Cave reports that 90 percent of the population of India believes in the soul, while in the African country of Nigeria, the number is close to 100 percent. Cave concludes that "all in all, the overwhelming majority of the world's nearly seven billion inhabitants subscribe to this particular immortality narrative."[22]

A few years ago, I became interested in whether the students I teach at Rutgers University also believe in the soul. After all, their psychology classes are all premised on the conclusion that the mind is nothing but an abstract description of the physical workings of the brain. And that, I thought, might have an effect on soul beliefs. To find out, I decided to ask the students themselves by having them fill out a questionnaire they took as part of a class assignment. Participants were asked to answer the questions as honestly as possible. I made it clear that I was interested in what they *really* believed about the soul, and not what they would politely tell their grandmother at a family dinner. Students were asked to indicate whether they believed in the soul by picking a value on the 6-point scale shown in figure 2.1.

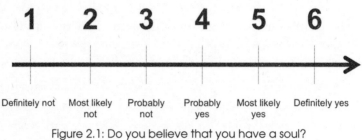

Figure 2.1: Do you believe that you have a soul?
Provide your answer by picking a value on the scale below.

What I found was rather surprising. Out of a group of close to two hundred students I taught in the fall of 2012, 80 percent picked a value on the soul belief scale between 4 and 6, indicating that they believed in the soul. Almost half of these "soul believers" were so confident in their belief that they selected the highest value on the scale, 6. In a series of follow-up questions, I asked the students to tell me more about their beliefs. They were given four choices to describe their soul:

(a) An immaterial entity separate from your body.
(b) A metaphor for the activity of your brain (in other words, your soul is purely physical).
(c) An outdated concept that is no longer relevant and plays no role in your functioning.
(d) Other.

Almost 3 in 4 soul believers, 73 percent, selected option (a). When I asked the students to tell me what functions their soul performed by selecting among a list of options, here's what they picked. The percentages represent the number of students who selected a particular option.

- A moral compass—71 percent
- Consciousness—67 percent
- Feelings—58 percent
- Free will—55 percent
- Personality—52 percent
- The ability to make decisions—48 percent
- The ability to fall in love—46 percent

I also asked participants about the relationship between their *soul* and their *mind*. They were given three choices:

(a) They are one and the same entity.

(b) They are separate entities but they overlap (in other words, your soul and your mind can sometimes perform the same function).

(c) They are completely separate entities that do not overlap (they perform entirely separate functions).

A majority of soul believers, 61 percent, selected option (b), indicating that they regard their soul as separate from their minds.

So much for the effect of psychology courses on soul beliefs! To make sure that these data weren't a fluke, I decided to repeat the experiment, this time by bringing students directly into the lab. In the fall of 2013, Hayley Fitzgerald, Sonia Skooglund, and I tested 220 Rutgers undergraduate students, using a computerized version of the soul questionnaire. We found that 88 percent of the participants were soul believers (they picked a value on the soul belief scale between 4 and 6). Of these students, 69 percent said that their soul was an immaterial entity separate from their body (option (a) in the first list above), and 86 percent claimed that their soul was immortal. Moreover, 69 percent of these soul believers indicated that consciousness was an aspect of the soul that would continue to exist in the afterlife. As before, the majority of students, 71 percent, regarded their soul as separate from their minds (option (b) in the second list above).

In response to questions about the role played by their souls, here's what these students indicated.

- A moral compass—74 percent
- Consciousness—68 percent
- Feelings—67 percent
- The ability to fall in love—67 percent
- Free will—64 percent
- Personality—61 percent
- The ability to make decisions—46 percent

We also wanted to know where students believed their soul came from, and found that 73 percent said that God gave it to them.

Let us take stock. The data from national polls combined with the more detailed studies we conducted with Rutgers students all point to the same conclusion. Most people in America today believe in a soul that has three core properties:

1. The soul is believed to be *immaterial*, or *nonphysical*, and therefore distinct from the body.
2. The soul is believed to be *psychologically potent*. It is what gives people their psychological powers, primarily consciousness and the ability to make moral decisions, but also free will, feelings, personality, and our broader ability to make decisions.
3. The soul is believed to be *immortal* and to carry consciousness into the afterlife.

Most people then are substance dualists. This is not to say that they believe in exactly the same soul Descartes wrote about in the seventeenth century, but it is pretty close. Nor would we want to conclude that people collapse the notion of personhood into a simple soul-body dichotomy. If we dug deeper, we might very well find that the notion of personhood was even more fragmented. But at the very least, it is safe to conclude that most people do not have a monistic view of personhood and that they believe in a form of substance dualism. This gives us a clear definition of the soul. This is the soul that this book is about. From now on, when I talk about *the soul*, I will be referring to the soul defined in points 1–3 above.

As I said earlier, people often use the word *soul* metaphorically, without any metaphysical connotation. In his book *I am a Strange Loop*, cognitive scientist Douglass Hofstadter writes about people's *souls* and the *little souls* of other animals, but he makes it abundantly clear that these are just figures of speech and that he is not a dualist.[23] Like most other cognitive scientists, Hofstadter is a committed materialist and his book is an attempt to make sense of consciousness in purely physical terms. The metaphorical use of the word soul is also beautifully captured in "Soul Talk," a piece that philosopher Stephen Asma wrote for the *Chronicle of Higher Education*.[24] Asma explains that comments like "She is an old soul," "He is my soul mate," or "James Brown has soul" are not factual claims about the world, but rather, that they express emotional attitudes on the part of the speaker. This is all very well, and I myself sometimes use the word soul metaphorically, but this is not what I will be concerned with in this book. Metaphors aside, I am interested in a much more fundamental question: Do people *really* have immortal souls? This is what we are going to try to find out.

MINORITY REPORT

As we discovered earlier, soul beliefs have been extremely common throughout history and up to the present day. There is one notable exception to this pattern however. As linguist Charles Hockett (1916–2000) remarked, "some very tiny segments of Western society reject the notion altogether."[25] These recalcitrant minds are today's scientists. In *The Soul Hypothesis*, Baker and Goetz observe that while soul beliefs have remained very common among the masses, they have all but vanished in scientific circles, a trend that started in the twentieth century and continued to accelerate in the early twenty-first century. Here too, we can find corroborating evidence. A 1998 article titled "Leading Scientists Still Reject God," published in the journal *Nature,* reports that 35 percent of elite scientists (members of the National Academy of Sciences) believed in human immortality in 1914, but that the number dropped to 18 percent in 1933, and a mere 7.9 percent in 1998[26]—exactly the trend described by Baker and Goetz.

The view that now dominates the sciences is materialism. It is Francis Crick's astonishing hypothesis. As we saw earlier, the idea isn't new and can be traced back to the medical observations of Egyptian surgeons more than four thousand years ago and to the ideas of Leucippus, Democritus, and Epicurus, the early Greek atomists. For most of history, materialism was the minority view, but the tides have now turned, and it has become the mainstream position in the sciences. In his 1991 book *Consciousness Explained*, Daniel Dennett offers the following assessment:

> The prevailing wisdom, variously expressed and argued for, is *materi-alism*: there is only one sort of stuff, namely matter—the physical stuff of physics, chemistry, and physiology—and the mind is somehow nothing but a physical phenomenon. In short, the mind is the brain. According to materialists, we can (in principle!) account for very mental phenomenon using the same physical principles, laws and raw materials that suffice to explain radioactivity, continental drift, photosynthesis, reproduction, nutrition, and growth.[27]

Other prominent public intellectuals, like psychologist Steven Pinker, biologist Richard Dawkins, and physicist Victor J. Stenger, reach the same conclusion. At the beginning of his book *The Blank Slate*, Pinker describes three myths that he believes must be abandoned. One of them is dualism, the doctrine of "the ghost in the machine" (we'll come back to the origins of that phrase in chapter 7). In the preface to his book *Physics and Psychics*, Victor Stenger summarizes the situation with a simple equation: $\Psi = \Phi$. The first symbol is the Greek letter *psi*, which stands for *psyche*, the world of mind and soul. The second symbol is the Greek letter *phi*, representing the material world of physics. Stenger's equation tells us that mental phenomena *are* material phenomena and gives Dennett's assessment a more compact formulation. Even in religious quarters, people seem to have noticed that the tides have turned and that dualism is under attack. The back cover of a 2004 book titled *What About the Soul? Neuroscience and Christian Anthropology*, warns its readers about the impending collision between dualism and modern science:

Everyone knows about the rocky relationship between science and theology brought about by the revolutionary proposals of Copernicus and Darwin. Fewer people know about an equally revolutionary scientific innovation that is currently under way among neurobiologists. This revolution in brain research has completely rewritten our understanding of who we are. It poses fundamental challenges to traditional Christian theology. *According to the scientific worldview that now dominates, it is no longer necessary to speak of a soul or spirit as distinct from the functions of the brain.* (Italics added.)[28]

At the end of his book *Descartes' Baby*, Paul Bloom asks whether our propensity to divide the world into body and soul has any basis in reality. Since we are fairly certain that bodies exist, what Bloom is really asking is whether souls are real too. Bloom, like most of his fellow scientists, is a materialist, and he believes in the physical basis of mind. But he does not underestimate the magnitude of the clash between the scientific view of personhood and the popular one. People may well learn to live with the fact that our intuitions about the soul are mistaken, Bloom tells us, but it won't be easy. One way to avoid this confrontation is to find refuge in the idea that science and the soul occupy nonoverlapping magisteria, to use Stephen J. Gould's famous phrase. Most people believe in the reality of the soul, but many also insist that the soul is *only* a philosophical, theological, or metaphysical claim that lies outside the scope of science and therefore cannot be affected by discoveries in physics, biology, or neuroscience.

THE SOUL *IS* A SCIENTIFIC HYPOTHESIS

One of the questions we asked the students in our soul-searching experiment was this:

Do you believe that the question of whether human beings have a soul is a question that can be answered objectively—at least in principle—and thus one that can be answered using scientific means?

Almost three quarters of soul believers, 74 percent, answered negatively. I beg to differ. Maintaining that the soul plays an active role in our psychological functioning, that it can operate independently from the body, and then trying to argue that these claims are not scientific is a clear case of doublespeak. This reminds me of the approach taken by Senator Jon Kyl after he publicly asserted that well over 90 percent of what Planned Parenthood does is perform abortions. When CNN journalists tried to contact the good Senator to point out that his numbers were . . . well, a little off, his office replied that the Senator's remark was "not intended to be a factual statement."[29]

Writing in 1935, the great Bertrand Russell already pointed out that soul claims represent a perfectly legitimate scientific hypothesis. In his classic essay *Religion and Science*, Russell explained that "There is, indeed, one line of argument in favour of survival after death which is, at least in intention, completely scientific. . . . It is clear that there could be evidence which would convince reasonable men."[30] What kind of evidence did Russell have in mind? Dualists insist that the soul, because it is distinct from the body, can function independently from it. This leads to the prediction that it should be possible, at least in principle, to communicate with the dead. Think of Aunt Emma's ashes in that urn on the fireplace mantle. Now if someone could establish meaningful communication with Aunt Emma, under controlled experimental conditions, then we would have the kind of evidence that Russell believed would convince reasonable men.

It so happens that the world doesn't seem to work this way, but it very well could have. If you and I were living in ancient Egypt five thousand years ago and you told me that we would one day see a man walk on the moon, I would have laughed in your face. If you told me that you talked to your dead mother the night before, I would have asked you for the details. We happen to live in a world in which men can walk on the moon but cannot talk to their dead mothers. If it had been the other way around, the world would have made just as much sense to us. Evidence compelled us to believe that men can walk on the moon (unless you believe in conspiracy theories), but it could easily have convinced us, if it were available, that it is possible to communicate with the dead. Soul beliefs, just like beliefs in men who can walk on the moon, are scientific hypotheses.

Here's another way to reach the same conclusion. Souls were conjured up by our ancestors to make sense of phenomena such as life and mind, two aspects of the natural world that scientists in the biological and cognitive sciences are actively studying today. Consequently, any claim that life or mind involves an immaterial soul-substance very explicitly encroaches on the turf of science. What's more, maintaining that human beings have souls amounts to making *a series of claims* relevant to different scientific disciplines. For one thing, the idea of an immaterial substance that can interact with our body to make us do the things that we do—act morally, feel sad or elated, or jump up and down on Oprah Winfrey's couch Tom Cruise–style—is a claim about physics. Physics is in the business of studying the effect of any force, field, or substance that can interact with material objects. Yes, I know, the soul is supposed to be *immaterial* (we'll come back to that later). So what? Newton proposed an invisible gravitational force that was every bit as "immaterial" to the Cartesians as their soul substance.[31] Did that prevent scientists from studying the effects of Newton's "immaterial" force?

The soul is not just a physical hypothesis, however. It is also a biological, psychological, and neuroscientific claim. We know that human beings evolved from more modest life forms—we were not planted on earth ten thousand years ago.[32] This means that the soul hypothesis is also a claim about biology. At what point in the unbroken physical chain between primitive life forms and human beings did souls get added to the mix and why? And if the soul gives us mental faculties such as consciousness, free will, or a moral compass, as most people today seem to believe, the same logic applies to the fields of psychology and neuroscience. Because belief in an immaterial soul represents a cluster of scientific hypotheses about physics, biology, and the sciences of the mind, determining whether we have souls is as objective a quest as answering questions about the origin of species or the age of the universe. The soul hypothesis is no different from earlier scientific claims about mysterious substances like *phlogiston*, the firelike element that was supposed to be released during combustion; the *luminiferous ether*, which was believed to be the medium through which light propagates; or the *life force* of the vitalists.

I am certainly not the first one to propose that supernatural claims can be investigated scientifically. The general line of reasoning I am adopting here has recently been championed by Victor J. Stenger and Richard Dawkins in their respective bestselling books *God: The Failed Hypothesis* and *The God Delusion*. Among the general public, there is often a nagging perception that science is dogmatically naturalistic and that it will not even admit of the possibility of supernatural phenomena. This was brought home to me by a comment I received from a student after I mentioned the Ouija board in one of my lectures. The student complained that scientists are stubbornly narrow-minded in their refusal to believe in the possibility of a spirit world. She was a Ouija board enthusiast who was convinced of the otherworldly nature of her communications. But the idea that science is narrow-minded rests on a misunderstanding. Scientists would be the first ones to embrace the existence of "supernatural" phenomena, including Ouija board spirits, if they could find credible evidence supporting such conclusions (more on this in chapter 3).

Those who claim that science is narrow-minded in the way I just described often conflate two important notions: *methodological naturalism* and *metaphysical naturalism*. Methodological naturalism is simply a commitment to the pursuit of objective knowledge through the scientific method. Saying that working scientists adhere to the tenets of methodological naturalism is like saying that painters use paint, brushes, and ladders to do their work. All painters use these things and all scientists, by definition, are methodological naturalists. Metaphysical naturalism is the much stronger claim that reality is *exclusively* composed of the naturalistic entities postulated by science (atoms, molecules, genes, and so on). Thus, whereas methodological naturalism makes no metaphysical claim about the ultimate nature of reality, and is therefore open to the possibility of supernatural phenomena, metaphysical naturalism closes the door on the supernatural as a matter of principle. I should add that there is nothing about the practice of normal science that forces scientists to endorse metaphysical naturalism. As a methodological naturalist, I say let's not be dogmatic and let's give the soul hypothesis a chance.

The conclusion that soul claims are scientific claims also follows from what the New Dualists themselves tell us in their books. These authors often explicitly invoke the notion of "evidence" or "proof" to convince their readers that souls are real, and they use these terms with their intended scientific meaning. Take for example Gary Joseph's *Proof of the Afterlife*, Jeffrey Long and Paul Perry's *Evidence of the Afterlife*, Dinesh D'Souza's *Life after Death: The Evidence*, or, even more recently, Eben Alexander's *Proof of Heaven* (whether these books actually prove anything is a different question—one that we will turn to in chapter 3). D'Souza's book is a particularly good example because in it he explicitly argues that the methods and principles of science demonstrate the existence of the soul and the afterlife. In their book *The Soul Hypothesis*, New Dualists Mark Baker and Stewart Goetz are adamant about the scientific status of the soul hypothesis:

> There is nothing inherently anti-scientific about the Soul Hypothesis. . . . Of course, it is worth considering whether scientific research might show that the Soul Hypothesis is false. . . . But that is a very different matter from saying that it is intrinsically anti-scientific and hence entirely out of bounds.[33]

I agree with Baker and Goetz here. Of course we should be open to the possibility that the soul hypothesis might turn out to be true, and we should not prejudge the conclusion before having examined the evidence. However, as I will show you in the pages ahead, scientists have discovered that there is no credible evidence for the existence of the soul, that soul claims fly in the face of what we know about modern science, and that there is overwhelming evidence supporting materialism.

FRAMING THE SOUL HYPOTHESIS

I love science-fiction movies. Aside from the entertainment value, they provide an endless supply of mind-bending scenarios. When *The Terminator* came out in the 1980s, I was captivated by the idea that a robot could pass for a real person. The wild shooting scenes, stunning special effects, and memorable one-liners certainly contributed to the awe I experienced as a teenager. More than two decades later, as I began to think about writing this book, Arnold Schwarzenegger's character came to mind. The Terminator is virtually indistinguishable from a real human being (it even has bad breath we are told), and it appears to be sentient. If you watch the movie carefully, you will notice that the Terminator is conscious. The killer robot has a first-person perspective, courtesy of his anthropomorphizing creator. To a Terminator, the world looks kind of reddish and it is filled with lines of code and fragments of language constantly floating before its (his?) eyes. And yet, as we all know, the Terminator contains no ghostly substance, no vital essence, no spiritual vapor, no soul—it is just a machine.

The central question I want to ask in this book is whether we are biological versions of the Terminator. By this I certainly do not mean to imply that we were programmed to kill Sarah Connor. Rather, I want to ask whether Mother Nature used only raw ingredients like atoms, molecules, and cells to spawn our beloved species. If so, materialism is true, and life, mind, and consciousness are the natural product of purely physical forces, however difficult this may be for most people to believe. The contending view is that human beings are composed of two fundamentally different substances, body stuff and soul stuff. This is what most people in America today believe. On this view, we may have the body of Terminators, but we have the minds of angels.

Before we ask whether dualism or materialism is true, we need to consider a critique of materialism that will allow us to bring the two contending positions into sharper focus. Nobody in the sciences today is seriously arguing that the stuff that makes up our bodies doesn't exist or that physicists, who study the properties of the material world, are wasting their time (notwithstanding the small clique of naysayers). It is probably fair to say that idealism, the doctrine that the material world is merely a figment of our imagination, while very popular in movies like *The Matrix,* is not so popular among scientists. The critique of materialism I would like to consider is related to the expanding basis upon which the doctrine rests. One of the chief proponents of this view is linguist and philosopher Noam Chomsky. Chomsky is an enormously influential intellectual who revolutionized the field of linguistics in the middle of the twentieth century, and has also been an extraordinarily prolific writer on a wide range of topics. The Cartesians of the seventeenth century, Chomsky tells us, could formulate a coherent version of the mind-body problem because they had a fairly clear conception of the notion of body. Recall that Descartes's first substance, *res extensa*, was defined in terms of his intuitive contact mechanics. (Think of two billiard balls smacking each other.)

In 1687, a few decades after Descartes's death, all that changed with the publication of Isaac Newton's magisterial *Philosophiæ Naturalis Principia Mathematica* (Mathematical Principles of Natural Philosophy). Newton's new physics came to replace Descartes's because it provided a much better account of the behavior of bodies and planetary motion. What is often overlooked, however, is that Newton reintroduced the kind of "occult" forces that Descartes and the early proponents of the mechanical philosophy were trying so hard to eliminate from their new, "intuitive" scientific explanations. Much to his own dismay, Newton had indeed discovered with his gravitational principle that material objects possess the intuitively unimaginable property of being able to act on each other at a distance, instantaneously, and through a vacuum.[34]

The thrust of Chomsky's argument is that ever since Newton replaced Descartes's intuitive contact mechanics, we have had no definite notion of body, and hence of matter, because what we call *material* or *physical* corresponds to an open-ended set of propositions subject to change and new developments in the physical sciences. In Chomsky's own words:

> What is the concept of body that finally emerged? The answer is that there is no clear and definite concept of body. If the best theory of the material world that we can construct includes a variety of forces, particles that have no mass, and other entities that would have been offensive to the "scientific common sense" of the Cartesians, then so be it: We conclude that these are properties of the physical world, the world of body. The conclusions are tentative, as befits empirical hypotheses, but are not subject to criticism because they transcend some a priori conception of body. There is no longer any definite conception of body. Rather, the material world is whatever we discover it to be, with whatever properties it must be assumed to have for the purposes of explanatory theory.[35]

If the material world is "whatever we discover it to be," then everything that will be discovered by science, however counterintuitive, will become "material" or "physical" by definition. Herein lies a fundamental asymmetry between materialism and dualism. In spite of its open-ended nature, materialism is defined *positively*. When scientists tell us that the world contains forces, particles with no mass, and other such entities, they define these notions, often with the help of equations, and they show us how these ideas enter into explanatory theories. In sharp contrast, dualism is defined *negatively*. When dualists tell us that the soul is nonphysical or immaterial, they do not tell us what it is, they tell us what it is *not*. We will come back to this problem in chapter 4.

We are now in a position to formulate our competing dualistic and materialistic hypotheses.

> *Dualistic hypothesis*: Human beings are composed of a physical body and an immaterial, psychologically potent, and immortal soul (the soul we defined earlier). Body and soul are distinct entities and the soul can continue to exist and function independently from the body after we die.
>
> *Materialistic hypothesis*: The mind, the domain of the soul, cannot function separately from the body for the simple reason that our mental experiences are caused by physical activity in our brains. What we call mind is nothing but a description of the functioning of the brain at a certain level of abstraction. Body and mind are therefore two sides of the same coin.

Let us look at these two hypotheses a little more closely. The dualistic hypothesis asserts that the body and the soul/mind are two separate (albeit interacting) substances or mechanisms. The body and the soul/mind are therefore *detachable*, and the soul/mind can continue to exist and function without the body. By contrast, the materialistic hypothesis maintains that the mind *is a property of the body* and therefore that mind and body are not *detachable*. Returning to the distinction between methodological and metaphysical naturalism, we can see now that the notion of detachability, which differentiates the dualistic and materialistic hypotheses, is a methodological claim—not a metaphysical one. By that I mean that the question of detachability is a perfectly well-defined empirical hypothesis that could be tested and settled using the scientific method (as Bertrand Russell explained back in 1935 and as we saw with the example of Aunt Emma).

Let us now briefly consider what follows from my use of the term *materialistic*. First, Chomsky's broader concern over how we should define materialism *as a general thesis*, does not affect the *specific question* about the detachability of body and mind that differentiates our two competing hypotheses, and so it won't be a problem for us. All I am saying really is that if materialism is true, then mind and body are not *detachable*. What I mean by this is that the human mind simply cannot exist, and therefore cannot function, without the human body. The question of detachability *is* the question of the soul. This is the question that we will try to answer in this book. To the extent that there are good reasons to believe that mind and body are *not* detachable, a second question arises. If body and mind are two sides of the same coin, then how can we reduce the latter to the operation of the former? I'll let philosophers worry about this question. My job as a scientist—a methodological naturalist—is to show you that there are indeed good reasons to believe that body and mind are *not* detachable, as predicted by our materialistic hypothesis.

An analogy will help bring our two competing hypotheses to life. Think of a portable CD player. I know, smartphones are better and newer, but the analogy works best with slightly older technology. You put in the CD of your choice, plug in your headphones, hit play, and presto, you can start enjoying your favorite tunes. Now imagine that someone asked you where the music comes from. The answer is straightforward: the music comes from the operation of the physical device you call a CD player. Once you hit stop, or destroy the CD player (including the CD itself), the music it was once able to play will not continue to exist in some kind of CD heaven (or hell if the music was really bad). Of course, there might be other copies of the same CD, and the music might be stored in other ways. But that is beside the point. If the *only* way to listen to the music was to play the CD in your portable CD player, then once you hit stop or destroy the CD player, the music would be gone forever, because it was *produced* by the operation of the CD player, and does not exist independently from it. Replace the CD player with

your body and brain, the music with your mind, and you have the materialistic hypothesis.

Now imagine a portable radio set. You select your favorite station, plug
in your headphones, and presto, you can start enjoying your favorite tunes. If
someone asked where the music comes from, the answer in this case would
be that your radio set receives a signal, in the form of radio waves (a kind
of electromagnetic signal). Since the waves and the device you call your
radio set are separate entities, the signal would continue to exist even if you
turned off your radio set or destroyed it. Someone with a different radio
set, or perhaps even a different device, could potentially receive the signal
and continue to enjoy the music. Replace the radio set with your body and
brain and the radio waves with your mind, and you have the dualistic hypothesis. Of course the analogy isn't perfect, because electromagnetic waves are
physical, but you get the idea. This analogy suggests that dualists have two
options to cash out their detachability claim. The first would be to say that
the separate "soul signal" is perfectly naturalistic. In this view, the soul signal
would be to us today what electromagnetic waves would have been to the
Cartesians of the seventeenth century—utterly mystifying, but nevertheless
perfectly natural. The second option would be to claim that the separate soul
signal is utterly otherworldly, and that it will never, as a matter of principle,
be understood by naturalistic science. We'll come back to this in chapter 5.

To summarize, most people in the United States today, and many others
around the world, believe in a soul that is (a) immaterial and distinct from the
body, (b) psychologically potent, and (c) immortal. This hypothesis amounts
to the claim that the mind is detachable from the body and can continue to
function without it. This idea, in turn, is a perfectly respectable scientific
hypothesis that may or may not be true. This reasoning applies, I should add,
to the *natural* capacities of human beings. Let me explain what I mean. We all
know that people cannot naturally fly or breathe underwater the way birds
and fish can. But this does not mean that human beings cannot fly and breathe
underwater period. Aided by scuba diving equipment and while traveling on
board airplanes, it becomes possible for us to fly and breathe underwater. As
our understanding of ourselves and our technology continues to improve, it
may one day become possible to download a person's "mind" into a machine.

If so the mind and the body would become detachable with the help of the right technology (Think of the movie *Transcendence*, for example). But this is clearly a different question. What we want to know is whether the mind and the body are *naturally* detachable. Likewise, a Martian exploring our planet could ask whether people can *naturally* fly or breathe underwater. The answer in all three cases, as I will show you, is no (but I'm sure you already knew about flying and breathing underwater).

To return to the forensic analogy I introduced earlier, I have now described the case to you, and we are ready to hear both sides present their arguments, so that you, the jury, can make up your mind about the existence of the soul. Before the hearings can begin, however, I need to tell you about the rules of the game. The competing claims we will examine in the pages ahead are scientific hypotheses, and so in order to decide which of our two hypotheses is more likely to be true, we will need to carefully weigh the evidence supporting each hypothesis. Evaluating evidence, it turns out, can be tricky business, especially for those not trained in the ways of science. For this reason, we will spend some time in the next chapter familiarizing ourselves with the way professional scientists evaluate evidence.

Chapter 3

THE FIRST PRINCIPLE

The first principle is that you must not fool yourself—and you are the easiest person to fool.
—Richard Feynman, *Surely You're Joking, Mr. Feynman*, 1985

Three centuries before Richard Feynman gave us his first principle, Francis Bacon, a British philosopher often regarded as one of the fathers of the scientific revolution, reached his own conclusions about the art of not fooling oneself. A true visionary, Bacon anticipated the broad outlines of an entire subfield of modern psychology concerned with the way human beings reason and make decisions—a body of work for which psychologist Daniel Kahneman was awarded the 2002 Nobel Prize in Economics. Using the metaphor of a clear glass through which beams of light are reflected according to their true incidence, Bacon warned us that the human mind operates in a rather different manner, much like an enchanted glass, full of superstition and imposture. If we want to have any hope of truly understanding the world, the mind's distorting biases must be corrected, Bacon insisted. His proposed remedy was the burgeoning set of ideas that was about to crystallize into what we know today as the scientific method.

In his *Novum Organum Scientiarum*, published in 1620, Bacon introduced the principles underlying the new philosophy and provided a remarkably lucid account of the difficulties we face in trying to understand the objective world. These obstacles are what Bacon called Idols—what today we would call biases. Bacon identified four types of Idols: Idols of the Tribe, Idols of the Cave, Idols of the Marketplace, and Idols of the Theatre. Idols of the Tribe are universal human tendencies, such as seeing patterns and regularities where there are none, letting our desires and feelings color our view of reality, or blindly trusting our senses—which are of course limited and fallible. (Think of being fooled by a magician.) Idols of the Cave are the set of idiosyncratic biases that can be found in each of us by virtue of our particular training, professions, or experiences. (As the saying goes, if you are holding a hammer, the whole world looks like a nail.) Idols of the Marketplace correspond to the many ways in which language can create confusion and ambiguity and lead us to chase our own shadows. (Think of Bill Clinton's deliberations on what the meaning of "is" actually is.) Finally, Idols of the Theatre are dogmatic sets of beliefs that stand in the way of knowledge. (Think of Galileo and the Church.)

As we gear up to embark on our soul-searching mission, it will be useful to remind ourselves of the many ways our minds can err. After all, we want to know whether people *really* have souls, and so we should take every precaution to steer our ship away from the dangerous vicinity of Bacon's Idols. Professional scientists are trained to recognize these trappings, and we all strive to follow Feynman's first principle in our work. Outside the hallways of academia, awareness of the mind's imperfections is dimmer, and while most people certainly know that they can occasionally get fooled, they may not realize the full extent and power of the hidden forces lurking beyond the reach of conscious awareness. This is of course not meant to denigrate lay people, but simply to remind ourselves that not everyone is a professional scientist, and that science has important lessons to teach us all.

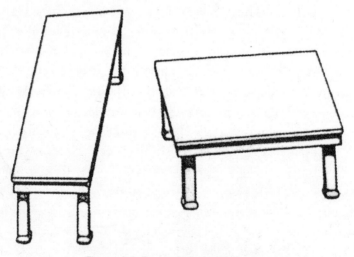

Figure 3.1: Roger Shepard's Tables.
Used with permission from Robert Shephard.

One of these lessons, I will argue—the central thesis of this book—is that belief in the soul is illusory. As an analogy, consider Roger Shepard's classic tabletop illusion, shown in figure 3.1.

Which of the two tables in that picture would you say is longer? Go ahead, consult your visual intuitions. Like the rest of us, I am sure that you will "see" that the table on the left is longer than the one on the right. Now grab a ruler and measure the length of each table. Surprised? You can make a photocopy of this page, cut out the two tabletops, and you will find by super-imposing one on top of the other that they are identical in length. Pretty neat, isn't it? Now that you have convinced yourself that the tabletops are identical, look at them again. Do you now see them as identical in length? No?! So not only are your visual intuitions objectively unreliable, they are also incredibly stubborn. You now *know* that the two tabletops are identical, and yet you continue to see one as being longer than the other!

Shepard's tabletop illusion demonstrates the power of Bacon's Idol of the Tribe. It pits our first-person perspective—what our senses subjectively reveal to us—against the third-person perspective—what the result of objective measurement demonstrates. The story of the demise of the soul, to a large extent, reflects the triumph of the third-person perspective over its subjective, first-person counterpart. But there is more to Shepard's illu-sion than meets the eye. The analogy contains two additional virtues. After this brief demonstration, I do not know anyone who would continue to insist that one tabletop *really* is longer than the other because of the way they look. All of us then, regardless of what we happen to believe about the soul, fully grasp the critical conclusion that reason must triumph over feelings and impressions if we want to understand how the world works.

Some people, among them scientists, take this general conclusion to heart, and apply it unabashedly to all aspects of the world. Others take a more opportunistic approach to the truth and only give in to reason when it proves convenient for them to do so; otherwise, for better or for worse, they cling to what feels right. The analogy's second virtue is that it reveals to us how stubborn first-person impressions can be in the face of objective evidence. Like Shepard's tabletop illusion, the soul illusion is a particularly stubborn one. (I'll try to explain why in chapter 7.) To dispel the soul illusion, I will show you in the next three chapters that we know too much about biology, physics, psychology, and neuroscience to let the illusionist— our own brains in this case—fool us into believing that we have a soul. The trick, however, is one of nature's most sophisticated, and it is so polished and compelling that it has fooled people for millennia. Before we start unraveling the soul illusion, we will first need to come face-to-face with Bacon's Idols, learn to tame them, and make sure we can keep them at bay as we proceed with our detective work. To prepare the terrain for our investigation, let me begin by telling you the story of Facilitated Communication, another spectacular demonstration of the power of Bacon's Idols.

TALK TO THE LETTER BOARD

When Douglas Biklen returned from a trip to Australia in 1989, he brought back with him a revolutionary idea. For thousands of families across America whose lives had been turned upside down by autism, a mysterious condition that cripples the mind, hope was on the way. Biklen had found a cure. He announced that contrary to received wisdom, nonverbal individuals with autism in fact possessed minds that functioned like yours or mine, but that they were trapped inside bodies they could not fully control. And now thanks to Biklen's brainchild, a new technique called Facilitated Communication (FC), all that was about to change. With the help of a facilitator to guide their hand, and a letter board on which they could type their thoughts, nonverbal autistic individuals were at last able to break free from their prison of silence. Jeff Powel, a student at Baker High School in Syracuse, New York, had long been regarded by his peers and teachers as someone who was profoundly disabled. But thanks to FC, Jeff's true self had finally found a voice. Jeff was in fact an academically gifted student who wrote poetry for the school yearbook and became a local celebrity.[1]

Stories like Jeff's began to multiply, and the demand for FC spread like wildfire. Young children with abysmal IQs could now read and write. Teenagers who had only been able to utter a handful of intelligible words were now solving math problems that baffled their parents. An entire group of people was on the brink of being redefined. As the demand to train facilitators skyrocketed, conferences were organized to promote the miracle technique. New schools, like the one in Syracuse, began to open all around the country. Thanks to Douglas Biklen and FC, thousands of people who were once regarded as hopelessly disabled had finally found a voice. It was hard not to be swept up by the wave of hope and optimism that took over the country.

Nevertheless, there were skeptics. One of them was Dr. Howard Shane at Boston Children's Hospital, who ran a center dedicated to finding technological solutions to allow disabled individuals to communicate independently. Shane, who worked with patients suffering from cerebral palsy, was able to capitalize on the slightest ability to move—the blinking of an eye, the twitching of a muscle—to control computer-based technology and allow profoundly disabled individuals to express themselves independently. With all the technology available, he asked, why do autistic individuals, who are not paralyzed, need people to hold their hands?

When Gerry Gherardi, a pharmacist from New Hampshire, arrived home from work one evening, his wife Cathy rushed out of the house to warn him that he was not allowed in and that there was a warrant out for his arrest. There were allegations that Gerry had sexually abused his seventeen-year-old autistic son, Matthew.[2] To make things worse, the author of the allegations was Matthew himself, who had been using FC at school. With the help of his facilitator, Matthew had typed the incriminating messages. Gerry Gherardi was forced to spend time away from his home, devastated by the words that came from the letter board. There was no other evidence that Gerry had abused his son, and yet everybody took the messages typed by Matthew and his facilitator at face value.

Similar cases of alleged sexual abuse surfaced in California, Texas, Georgia, Indiana, Oklahoma, and New York. A seventeen-year-old autistic girl from Maine named Betsy Wheaton used FC to write that her entire family was sexually abusing her. Betsy was immediately put in foster care under the supervision of attorney Phil Worden, who served as her legal guardian. Worden understood that the stakes were unusually high in this case and that before reaching a verdict it was crucial for the court to determine whether the communications were coming from Betsy or from her facilitator. Dr. Howard Shane, our skeptic from Boston's Children Hospital, was asked to evaluate Betsy.

Shane decided to test Betsy and her facilitator using a standard blind procedure, in order to rigorously determine who was the author of the typed messages. Imagine that you were to show Betsy and her facilitator the same picture and that you asked Betsy to type what she saw as her facilitator guides her hand. If both Betsy and her facilitator saw a picture of a dog, we would expect the typed message to come out as D-O-G. But now imagine what would happen if Betsy and her facilitator were shown different pictures (without knowing that they each saw something different). What if Betsy saw a picture of a dog, as before, but her facilitator saw a picture of a shoe?

If Betsy is really the author of the communications, the typed message should be D-O-G—what Betsy saw. However, if the communications come from Betsy's facilitator, the message should be S-H-O-E—what the facilitator saw. The results of Shane's simple experiment were devastatingly clear: not once did the word corresponding to the picture that Betsy had seen end up being typed. Instead, the names of the pictures that Betsy's facilitator saw emerged from the letter board. What Shane unequivocally demonstrated is that Betsy was not the author of her alleged communications—her facilitator was. Subsequent studies published in peer-reviewed journals involving large numbers of autistic individuals and hundreds of trials replicated Shane's basic finding. In spite of all the hype and hope surrounding the new technique, FC simply didn't work.[3]

ELEMENTARY, MY DEAR WATSON

Hindsight is always 20/20, and upon reading the story of FC, it is difficult to believe that so many people had the wool pulled over their eyes for so long. Why didn't anyone think of testing the technique before it was sold to the general public as a miracle cure for autism? Fortunately, every cloud has a silver lining, and the story of FC also illustrates the perils of Bacon's Idols. In their willingness to see a causal relation where none existed, far too many people, buoyed by hope and optimism, were blinded by the Idol of the Tribe and were led to draw strong conclusions on the basis of unreliable evidence. As we consider the notion of evidence, we can also see Bacon's Idol of the Marketplace rearing its ugly head. In English, the word *evidence* is often used with different meanings in ordinary parlance and in scientific discourse. All decisive evidence is evidence, but not everything that people call evidence counts as decisive evidence. And if we want to avoid fooling ourselves, decisive evidence is what we need to learn to recognize.

Given what you now know about Douglas Biklen's technique, imagine that one of your friends—let's call him Wayne—was to praise the merits of FC and recommend its use to families of autistic children. If you asked Wayne to tell you how he knows that FC works, he would gleefully point to the messages emerging from the letter boards. He might even tell you that he personally knows families who use FC with their children and that the technique does work. But we all know that Wayne would be fooling himself. What he regards as "evidence" in this situation is what scientists would call *anecdotal evidence*, a weak form of evidence that doesn't count as real evidence because it isn't decisive. Wayne's anecdotal evidence is certainly evidence of something (that messages were typed), but it is not decisive evidence that the autistic individuals were the authors of the messages.

The problem here is that if the autistic individuals were indeed the authors of the messages, you would observe precisely what you are observing. If, on the other hand, the facilitators were the authors of the messages, you would also observe precisely what you are observing. So how could you possibly know who's really doing what? One would think that this simple observation would be enough to deflate Wayne's enthusiasm. But as soon as you were done explaining to him that anecdotal evidence isn't reliable, Wayne would give you more "evidence" that his interpretation must be the correct one. In order to believe that the facilitators are the authors of the messages, Wayne would tell you, you need to make the counterintuitive assumption that they are influencing autistic individuals *unconsciously*. For who, apart from disgruntled conspiracy theorists, would be willing to believe that the facilitators all purposely put words in the mouths of nonverbal autistic individuals?

This is where training in experimental psychology comes in handy. Wayne's counterintuitive assumption is in fact a well-established scientific conclusion named after a famous horse. We owe the discovery of this effect to Wilhelm von Hosten, a late nineteenth-century mathematics teacher from Germany, and his horse Hans. Von Hosten, an amateur horse trainer, was convinced that he had managed to teach his horse Hans to solve arithmetic problems. Hans was trained to tap the number corresponding to the correct answer with his hoof. Von Hosten was so proud of their success that the pair started performing in front of live audiences throughout Germany. Predictably, Clever Hans, as the horse came to be known, attracted the attention of scientists, and a commission was formed to investigate the phenomenon. In 1904, the commission concluded that no fraud was involved in Hans's performance.

Psychologist Carl Stumpf, one of the members of the Hans Commission, was charged with the task of further investigating the horse's alleged abilities. After performing a number of controlled experiments, Stumpf discovered that Hans got the right answer only when he could see the questioner and the questioner knew the right answer. This led Stumpf to examine the questioner's behavior and notice that as Hans's taps got closer to the correct answer, tension became manifest in the questioner's posture and facial expression, only to be released right after Hans made his final tap—the one corresponding to the right answer. Hans didn't possess extraordinary mathematical powers after all, but he was very clever indeed and was able to pick up on subtle, unconscious cues produced by the questioner. The Clever Hans effect was an important discovery, and we owe double-blind experimental protocols to Stumpf's brilliant insight.[4]

Returning to your friend Wayne—you do not need to be a conspiracy theorist to believe that facilitators can unconsciously influence the typing of autistic individuals. Clever Hans's questioners were not cognizant of the fact that their posture and facial expressions gave the horse what he needed to arrive at the right answer. Likewise, facilitators do not consciously realize that they are influencing the typing. This phenomenon is similar to what happens with table turning, a very popular practice in the nineteenth century, or with its counterpart on today's college campuses, the Ouija board. Table turning, as the name suggests, involves a group of people placing their hands on a table, invoking their favorite spirits, and waiting for the mischievous balls of ectoplasm to start moving the table. Tables do turn, and spirits were held accountable, at least until English scientist Michael Faraday conclusively demonstrated that the tables were not set in motion by immaterial forces but by the very material effects of unconscious muscular action called ideo-motor movements.[5]

I am telling you about Facilitated Communication, von Hosten's horse, and Michael Faraday's ghost-busting act for a reason. My plan in the next chapter is to show you that the evidence for the soul invoked by the New Dualists, like the purported evidence for FC, does not count as decisive evidence. What I am doing right now is getting you to appreciate important distinctions that any reasonable person needs to make. By planting the seeds of these ideas in your mind now, I am relying on an effect discussed by Amos Tversky and Daniel Kahneman, our 2002 Nobel laureate, called the Availability Heuristic. The idea behind Availability is that when people make judgments about the probability of an event, they unconsciously base those judgments on how easy it is to think about examples of similar events. By showing you the importance of certain conclusions now, I will make it easier for you to see their relevance and apply them when you need to do so in the next chapter.

PROS VS. JOES

A few years ago, Spike TV launched a new kind of reality game show called *Pros vs. Joes*, in which opposing teams of amateur athletes (the Joes) and retired professional athletes (the Pros) competed. The idea itself is hilarious because we all know what would happen in a real face-off between, say, a professional tennis player and a regular person who enjoys a good game of tennis on the weekend. How could someone with average athletic abilities and even a moderate amount of experience possibly compete against someone with exceptional athletic abilities who has dedicated his entire life to the sport? When it comes to science, there are Pros and Joes too. The Joes are most people, and the Pros are those who have dedicated their lives to the profession; gone to graduate school; received postdoctoral training; learned to master complex statistical, computational, or mathematical techniques; managed to get their ideas published in peer-reviewed journals; received competitive funding from federal agencies; and been awarded tenure at research universities. The difference between a professional biologist with an established academic career and an average person who may have taken a few biology courses in college would be equivalent to the difference between an internationally ranked player and a weekend tennis enthusiast.

Not surprisingly, the people who decide whether there is evidence that tobacco causes cancer, whether global warming is real, whether vaccination causes autism, or whether the human mind works this way or that way, are trained scientists in the relevant disciplines. The opinions of nonscientists in settling scientific matters simply do not count. It would make about as much sense to let nonscientists weigh in on scientific issues as it would to give me a spot to play against Roger Federer in the US Open finals. If what I am telling you reeks of elitism, consider the following proposition: The next time someone in your family requires surgery, I'll offer my services and tell you that I can remove your daughter's appendix. I'll of course admit to having no medical training whatsoever, but I will add that I have always been a big fan of the show *ER*. If you need to fly somewhere, I'll volunteer to be your pilot, and assure you that I have a very good flight simulator on my new PlayStation system at home. Would you take me up on those well-intentioned offers if the lives of your loved ones were at stake?

Sadly, the difference between serious professional science and pseudo-scientific fluff is often blurred today, courtesy of the media and their obsession with "fairness." Take for example the case of actress Jenny McCarthy, a fervent believer in the thoroughly discredited idea that vaccination causes autism.[6] In 2007, McCarthy was given a national platform to broadcast her views on autism when she appeared on the *Oprah Winfrey Show*. McCarthy dismissed the opinions of medical professionals and told Oprah that her decisions were informed by "a little voice" and her "mommy instincts." She also appeared on *Larry King Live* and *Good Morning America*. Amazingly, Larry King had her debate a doctor, as though McCarthy's views were equivalent to those of a trained professional representing an entire field. When Oprah asked her about her credentials, McCarthy replied that she graduated from "the University of Google." In July 2013, the *Nation* published a scathing piece called "Jenny McCarthy's Vaccination Fear-Mongering and the Cult of False Equivalence." The title says it all.

As we learned in the previous chapter, the claim that human beings are ensouled is a scientific one. Consequently, it is up to professional scientists to decide whether there is evidence supporting such a claim. The cult of false equivalence would have us believe that the opinions of celebrities, priests, and politicians count when it comes to the soul. But why should that be so? What could these people possibly know about human biology, psychology, or neurophysiology that professional scientists do not know? If priests and politicians want to make claims about the nature and functioning of the human mind, then they clearly missed their calling—they should be psychologists or neuroscientists. The consensus in the sciences of the mind, as we discovered in chapter 2, is that the soul is a figment of our imagination. I certainly do not want to imply that the scientific consensus is infallible, or try to sell you a version of the argument from authority. The reason we should trust science is not because its practitioners use Latin words and have advanced degrees, but critically, because the currency in science is evidence. Generally speaking, it is evidence that compels scientists to converge on a particular set of conclusions. We know that species are not immutable, that smoking causes cancer, and that anthropogenic climate change is real, because the best scientific evidence has convinced the overwhelming majority of scientists in the relevant fields that these conclusions are true. What I will show you in the next two chapters is that it is also evidence and reason that have led mainstream scientists to abandon the soul.

This is not to say that the scientific establishment never gets things wrong. However, at any given point in time, the scientific consensus remains the best measure of truth at our collective disposal. If what physicists tell us about the physical world turns out to be wrong, who but future generations of physicists would be in a position to know better? The Pros remain the Pros and the Joes remain the Joes. The conclusion that science isn't perfect shouldn't be an open door to substitute evidence with "little voices," "mommy instincts," or other forms of irrational conjecture. What is needed to correct faulty scientific conclusions is better science, not uninformed speculation. As Albert Einstein reminded us, our science may be primitive and childlike, but it is the most precious thing we have.

Mention of the scientific consensus inevitably brings up the lure of what I call the Galileo syndrome, a piece of warped logic passionately articulated by Governor of Texas Rick Perry in front of a national audience during the run-up to the 2012 presidential election. When asked about his views on global warming, Perry declared that the science wasn't settled and that he therefore would not want to jeopardize the economy on the basis of what a group of scientists believed. Perry then invoked Galileo to make the point, I presume, that the consensus position can be wrong. We'll set aside the fact that Galileo was embroiled with the church in the early seventeenth century. For now, we will focus on the logic of the Galileo syndrome.

NAYSAYERS AND THE CURIOUS CASE OF DARYL BEM

The fact that scientists are more immune to the influence of Bacon's Idols than other groups of people does not imply that all members of the scientific community are impervious to flawed thinking. Even when scientific conclusions have acquired the status of factual knowledge, it is always possible to find people, sometimes even scientists, who seem to have a problem accepting reality. You would think that in our age of air travel and satellite communication, every single person with a modicum of education would accept the conclusion that the earth is not flat. But this would be counting without naysayers (to use a polite term).

I first heard about the Flat Earth Society while attending a fascinating lecture on different strands of creationism in the United States. This most peculiar of societies was founded in the late 1950s by Englishman Samuel Shenton. Californian Charles Johnson took the reins until his death in 2001, and the society made a surprising comeback in 2004, under the leadership of Daniel Shenton, an American expat who now lives in London. In February 2010, the British newspaper the *Guardian* ran an amusing piece on Shenton and the Flat Earth Society called "The Earth Is Flat? What Planet Is He On?"

The point of this brief excursion into bizarre-land is to talk about people who reject the scientific consensus even in the face of astronomical odds. Fortunately, in many cases, it is easy enough for anyone to see that the naysayers suffer from a severe form of stick-your-head-in-the-sand-ism. On other occasions, the naysayers are more refined, and they use their scientific credentials to try to convince the general public that the world is indeed full of superstition and imposture. It is this latter brand of scientifically savvy naysayers that I want to warn you about. Their conclusions are every bit as outrageous as those of their fellow flat-earthers, but the claims come from bona fide scientists, armed with doctorates in the relevant disciplines and genuine academic credentials.

Take for example Dr. Daryl Bem, Professor Emeritus of Psychology at Cornell University. In 2011, Bem published a paper titled "Feeling the Future: Experimental Evidence for Anomalous Retroactive Influences on Cognition and Affect," in the peer-reviewed *Journal of Personality and Social Psychology*. In his paper, Bem claimed to have uncovered evidence for *psi*, or what is sometimes called *precognition*. In plain English, Bem claimed to have discovered that ordinary people have psychic powers and that they can predict the future. In one of his experiments, Bem showed subjects two curtains displayed on a computer monitor and told them that an image would appear behind one of the curtains. Participants had to guess which one.

When shown most images, participants selected the correct curtain 49.8 percent of the time, a result that did not differ from chance performance. But when the images were erotic in nature, participants selected the right curtain 53.1 percent of the time, a statistically significant deviation from chance according to Bem. Before it was even published, "Feeling the Future" was discussed in the media, and Bem appeared as a guest on the *Colbert Report*. True to form, Steven Colbert asked Bem why psychic abilities seemed to only work with the sexy bits. Bem replied that he had no idea. To his credit, Colbert also pointed out that Bem's conclusions were highly controversial, and he mentioned Ray Hyman, a professor of psychology at the University of Oregon, who concluded that Bem's methods were flawed, that his results were unreliable, and that his article was an embarrassment for the entire field.

Hyman is not the only scientist to reach the conclusion that Bem's "evidence" cannot be taken seriously. Since the publication of "Feeling the Future," detailed critiques of Bem's general methodology, the design of his experiments, and his use of statistics have been published in peer-reviewed journals and in popular scientific magazines.[7] Several attempts have also been made by other scientists to replicate Bem's results, and they all failed.[8] But we shouldn't be surprised that Bem has failed to convince the scientific community. After all, his claim that people possess psychic powers is by no means new and original. The idea has been around for hundreds, if not thousands, of years, and one could write entire books documenting previous refutations of extravagant claims such as Bem's. In fact, people do write such books, among them skeptic-in-chief Victor J. Stenger, who wrote an excellent book on this topic called *Physics and Psychics*. Carl Sagan's *The Demon Haunted World* is also well worth reading in this genre.

Another important consideration to bear in mind when pondering Bem's extraordinary claims is that they fly in the face of everything we know about physics. Theoretical physicist Michio Kaku explains why in his book *Physics of the Impossible*. Here's what Kaku writes about precognition:

> In summary, precognition is ruled out by Newtonian physics. The iron rule of cause and effect is never violated. In the quantum theory, new states of matter are possible, such as antimatter, which corresponds to matter going backward in time, but causality is not violated. In fact, in quantum theory, antimatter is essential to restoring causality. Tachyons at first seem to violate causality, but physicists believe that their true purpose was to set off the big bang and hence they are not observable anymore. Therefore precognition seems to be ruled out for the foreseeable future, making it a class III impossibility. It would set off a major shake-up in the very foundations of modern physics if precognition was ever proved in reproducible experiments.[9]

To anyone who owns a cell phone, one of the countless demonstrations of the success of modern physics, Kaku's comments should serve as a sobering reminder regarding the utter implausibility of Bem's fantastical claims. To make things worse for Bem, critics are not just waving their hands and screaming foul play only to scratch their heads when asked to tell you how and where Bem went wrong. Other scientists can pinpoint precise aspects of

Bem's methods and explain why they are flawed. A recent article published in the journal *Psychological Science* demonstrates how easy it is to obtain false positive results in psychological experiments—precisely the type of effect that critics argue Bem obtained in "Feeling the Future."[10]

If we were to follow Rick Perry's logic, we would conclude that ESP is real because its validity has been demonstrated by a legitimate scientist, no doubt a modern-day Galileo, who published his conclusions in a peer-reviewed journal. Take that, you narrow-minded skeptic! The crucial piece of information that is invariably omitted in such stories is that while Daryl Bem's paper did manage to slip through the cracks of the peer-review process, it didn't manage to convince the scientific community. Hence the importance of the scientific consensus. In the case of ESP, the consensus remains that there is no credible evidence that the phenomenon is real. Since Bem failed to convince the rest of his scientific peers, he simply has no case. And people like Rick Perry, or your friend Wayne, who are scientific Joes, have no basis upon which to decide whether Bem is more likely to be right than the rest of his scientific peers. This is because the details of Bem's claims involve technical issues in experimental methodology and statistical analysis that people not trained in the sciences are ill equipped to understand. In the end, for every genuine scientific genius, there are countless pretenders whose extraordinary claims belong to the ash heap of history.

Those who are in the best position to evaluate claims like Bem's are scientific Pros. The existence of a published paper supporting your favorite extraordinary claim does not, in and of itself, provide decisive evidence in favor of that claim. Trade books are in many ways worse because they are not constrained by the peer-review process, which explains why they represent the weapon of choice in the arsenal of gurus and the scientific naysayers. Here's how the game is usually played: Since Galileo wannabes cannot convince the scientific community through publication in peer-reviewed journals, they bypass the scientific process altogether (as Douglas Biklen did with FC) and pitch their extraordinary ideas directly to the general public, all too often with astonishing success. I am of course writing a popular book myself, but it is merely to tell you about established scientific conclusions, not to peddle extraordinary claims.

In this regard, a number of books have surfaced recently, all making the case for the existence of the soul on the basis of what the authors allege to be scientific evidence. These books include *Life after Death: The Evidence* by conservative writer and Christian apologist Dinesh D'Souza, *Life after Death: The Burden of Proof* by New Age guru Deepak Chopra, *The Soul Hypothesis: Investigations into the Existence of the Soul* by linguist Mark Baker and philosopher Stewart Goetz, *The Spiritual Brain: A Neuroscientist's Case for the Existence of the Soul* by neuroscientist Mario Beauregard and journalist Denyse O'Leary, and *Proof of Heaven* by neurosurgeon Eben Alexander. The most commercially successful of these books, reported to have sold millions of copies, is Eben Alexander's *Proof of Heaven*. What seems to impress people is that Alexander is a neurosurgeon who taught at Harvard Medical School. With such credentials, many people believe, he's surely the next Galileo about to revolutionize the sciences of mind. Well, not really.

The New Dualists are in many ways worse than Daryl Bem. At least Bem tried to carry out controlled experiments to substantiate his extraordinary claims. The New Dualists instead choose the path of least resistance and pitch their ideas directly to the general public. Eben Alexander, for example, appeared on numerous TV shows and was touted as an expert whose views should be taken seriously, as though he was representative of the scientific establishment. This is of course a travesty. Alexander represents one of a handful of individuals in the neuroscientific community who actually believe in the soul and heaven, and he has been utterly unable to convince the rest of the scientific community that his claims have any merit. This is because of another feature of the scientific process, called the burden of proof.

RUSSELL'S FLYING TEAPOT

In the PBS documentary *Prisoners of Silence*, a thoughtful chronicle of the rise and fall of Facilitated Communication, Douglas Biklen is asked why he is unwilling to change his mind and remains committed to FC in the face of all the studies showing that the technique doesn't work. Biklen answers that it is very easy to fail to demonstrate something, and therefore that it almost doesn't matter how many studies fail to demonstrate the reality of FC. Biklen cleverly turns the tables on his critics, suggesting that there may be something wrong with their studies. This way of thinking betrays a common form of muddled reasoning characterized by a failure to recognize who actually bears the burden of proof. To see this, imagine that you received a letter summoning you to appear in court for an alleged speeding violation. When you get to court, the judge concedes that there is no positive evidence that you were speeding, but that to avoid being penalized, you have to prove that you didn't speed during your commute on that particular day. After reflecting for a few moments, you realize that you cannot possibly prove that you didn't speed, and so you tell the judge that you cannot disprove her claim. "Since you cannot disprove the claim," the judge proclaims, "I will conclude that you were indeed speeding. You now have two weeks to regularize your situation and pay your $250 ticket." Fortunately, the justice system doesn't work this way, and if it did it wouldn't last very long. Clearly, if someone accuses you of an offense or a crime, the burden of proof falls on *her* to demonstrate that you did indeed commit the crime—it doesn't fall on you to prove that you didn't.

The logic of this simple analogy extends far beyond the courtroom and carries with it important implications. The general rule at play here is that the very act of making a factual claim saddles you with the burden of proof. If you cannot provide evidence to substantiate your claim, then you simply do not have a case, and nobody should feel compelled to believe you. Of course, the nature of the claim matters and what is asserted is always interpreted against our background knowledge. If I were to claim that I can run a mile in under an hour, I wouldn't face a particularly heavy burden of proof given my age, my general physical condition, and what we know about human locomotion. But if I claimed that I can run a mile in under a minute, you should rightly demand that I produce evidence that I am capable of such a feat. If instead I told you, with a twinkle in my eye, that you cannot disprove my claim, you would be entitled to laugh in my face.

Returning to FC, the claim that nonverbal autistic individuals have minds that function in the same way as nonautistic individuals but who are trapped inside of bodies that they cannot fully control is precisely what needed to be demonstrated in the first place. Since Biklen's claim is a scientific hypothesis, the first step would have been to produce evidence capable of convincing professional scientists. Unfortunately, that critical first step was bypassed. Instead of first establishing its credibility in the court of science, FC was simply sold to the general public—and with devastating consequences. Failing to recognize who bears the burden of proof has further ramifications. There are many aspects of the world that science does not currently understand, an obvious fact that scientists routinely acknowledge. This creates gaps in our understanding of the world that believers in extraordinary claims are often eager to fill with their favorite "explanations." A classic illustration of this god-of-the-gaps argument comes from an exchange between Fox News's Bill O'Reilly and biologist Richard Dawkins. During the interview, Dawkins told O'Reilly that science didn't know how life emerged on earth. O'Reilly was quick to reply that until science can figure it out, he'd stick with Judeo-Christian philosophy. O'Reilly's logic is that if science cannot explain something, we can legitimately conclude that God did it.

As Dawkins tried to explain to O'Reilly, this form of reasoning reflects "a most extraordinary piece of warped logic."[11] There are countless myths and other logically possible stories that could explain the emergence of life on earth, including entirely naturalistic explanations. The fact that some aspects of nature remain unexplained does not lend credibility to a specific hypothesis among the myriad possible alternatives. Rather, each specific hypothesis bears its own burden of proof. Only after credible evidence has been given for a specific hypothesis can the case be made for a particular explanation. God does not win by default. If scientists in any field were to give talks at professional conferences in which they began by stating their favorite conclusions and then instead of presenting detailed evidence to support the validity of their claims simply challenged their audience to prove them wrong, they would be laughed off the stage.

One of the best-known expositions of the logic underlying the burden of proof is due to Bertrand Russell (1872–1970). Russell is known for his work with Alfred Whitehead on the foundations of mathematics, his best-selling *History of Western Philosophy*, his social and antiwar activism, his criticism of Christianity, and the Nobel Prize in literature he was awarded in

1950. In a 1952 essay titled "Is There a God?" Russell explains that it isn't the business of skeptics to disprove received dogma but rather that believers have to prove their claims. Russell asks us to imagine the claim that somewhere between the Earth and Mars there is a china teapot in orbit around the sun, too small to be revealed by the most powerful telescopes. The claim could thus not be disproved, but it would be ludicrous to accept it as true simply because it cannot be proved false. The burden of proof falls on the person claiming that such a celestial teapot exists. Failure to demonstrate the existence of the teapot renders the claim toothless and therefore not worthy of belief.

The soul hypothesis is a scientific claim about the detachability of mind and body and the existence of a mysterious substance powering our mental lives. It follows that anyone claiming that souls exist bears a burden of proof. What exactly is the evidence that should lead us to accept the existence of the soul? Before we consider the various lines of reasoning invoked by the New Dualists, I need to alert you to yet another danger posed by Bacon's Idols. Today, the province of the soul is the domain of the mind. In psychological matters, it seems that Pros and Joes contend on a level playing field. After all, both professional scientists and lay people have minds, and so, as Descartes would have reminded us, we all have privileged access to the content of our own minds. I know what it is like to be me from the inside and you don't. So why should we rely on science to tell us what is going on inside our own minds, when we are in a much better position to find out ourselves?

IT'S ALL IN YOUR HEAD

Think about your heart. You may be aware of its gentle beating and feel it accelerate in the presence of a loved one. You can detect your pulse by pressing your fingers on your wrist, and you can also tell that your heart is located in your chest. The information about the heart available to introspection is important, but if our understanding of this vital organ were limited to what our intuitions can deliver, we would know very little indeed. What about the shape and structure of your heart? How many parts does it have, and how do they relate to each other? What is the mechanism that allows your heart to beat in the first place? On these questions, our first-person intuitions remain in complete darkness. Of course, we know the answer to these questions, but it is not because we can intuit the relevant facts. Rather, we had to be told what the answers are or read about them. To be clear, I am asking you to forget about everything you've learned about your heart and to imagine that you could only know what introspection reveals to you about it.

This cardiac analogy illustrates a manifestation of the Idol of the Tribe that I call the first-person fallacy. Because we are sentient creatures and have access to an inner world of thoughts and sensations, it is easy to assume that our privileged access to this private world provides a reliable guide to our true nature and functioning. In the case of the heart, the fallacy is easy to recognize. A moment's reflection suffices to realize that the third-person perspective offered by science stands in an incomparably better position to tell us how the heart works, compared to our untutored, first-person intuitions. In the domain of the mind, the first-person fallacy requires more work to expose, and it will take us the remainder of this chapter and some additional considerations in the next one to fully come to grips with it. The objection we'll have to confront can be expressed in the following way: I have direct and private access to the content of my own thoughts and you're not me, dear scientist, so how could you and your third-person perspective possibly know better than me what makes me tick?

As a first step, I will show you that the rich and colorful tapestry of thoughts, feelings, impressions, and sensations that enter our steam of consciousness represent the product of an extraordinarily complex web of processes of which we have absolutely no awareness. The mind is like a gigantic iceberg—only the tip is accessible to conscious awareness. The bulk of the iceberg lies below the surface, and without the help of science we would be unaware of its existence and of the influence it exerts on us. Worse, our conscious mind often makes up stories about how it thinks it functions. In short, when it comes to understanding how the world works, our conscious minds and the intuitions to which they give rise, if stripped from what they have learned through education, are shallow, severely shortsighted, and prone to serious confabulation. Francis Bacon wasn't kidding. Of course, we are all very good at recognizing grandma's voice, we can tell chocolate from mustard, and we don't constantly walk into walls. My point is not that the conscious mind is useless—it obviously isn't—but simply that our untutored intuitions about how the world works are not reliable. This is why we need science.

To illustrate these conclusions, and begin eroding your conviction that you know yourself, let me tell you about the results of a fascinating set of experiments that involved literally splitting people's brains. You can think of your brain as two halves of a giant walnut, its two hemispheres connected by a dense bundle of nerve fibers known as the *corpus callosum*. In severe cases of epilepsy, a last-resort option consists of surgically severing the corpus callosum to prevent the electrical disturbance from traveling from one hemisphere to the other. Patients who undergo this procedure thus end up with a split brain, quite literally. Amazingly, the procedure produces little evidence that something so radical took place, and the patients report not feeling any different before and after the surgery.[12]

However, strange things begin to happen when split-brain patients are brought into the lab. This is what neuropsychologist Roger Sperry and his graduate student Michael Gazzaniga discovered in the 1960s. Their findings were so stunning that Sperry was awarded the 1981 Nobel Prize in Physiology/Medicine for his work with split-brain patients. The approach Sperry and Gazzaniga took isn't unlike Dr. Shane's, our skeptical doctor from Boston's Children Hospital. Shane's problem was that Betsy and her facilitator were always working in tandem, and so it was difficult to establish who was really doing what. Like Betsy and her facilitator, the two hemispheres of the brain also work in tandem, sharing information via the corpus callosum, and so it is not easy to understand exactly what each hemisphere does. But in split-brain patients, much less information is able to travel between the two disconnected hemispheres, offering researchers a rare opportunity to assess the role played by each hemisphere in various behavioral and perceptual tasks.

In their experiments, Sperry and Gazzaniga relied on the fact that information presented in the left visual field is processed by the right hemisphere of the brain and vice versa. They showed split-brain patients words that were briefly presented in the left or the right visual field. When the words appeared in the right visual field, the information was processed by the left hemisphere of the brain, and the patients could easily report the words they had seen. But when the words were presented in the left visual field, and were processed by the right hemisphere, the patients could no longer report what they had seen. Sperry and Gazzaniga had shown that only the left side of the brain can articulate speech.

This is where things get really interesting. Imagine presenting the word *hammer* to the left visual field of a split-brain patient and asking her to report what she saw. As before, the patient will tell you that she didn't see anything. If you now ask her to select with her left hand one of the objects on a tray hidden from view, she will pick up the hammer! The explanation for this surprising behavior is really nifty. The right hemisphere controls the left side of the body, including the left arm, and vice versa. The word *hammer* was presented to the patient's left visual field, and so the information was processed by the right hemisphere. The right hemisphere therefore knew that it had seen the word *hammer* even if the patient herself had no conscious awareness of it. Since the right hemisphere controls the left arm, it was able to tell the left hand to pick up the hammer. If the word flashed in the left visual field had been *screwdriver*, the patient's left hand would have grabbed the screwdriver from the tray.

This classic example from the annals of cognitive neuroscience vividly illustrates the role of *unconscious* mental information—the bulk of the iceberg in my earlier analogy. It is quite clear that the patient's conscious mind had no idea about the word that was presented to her left visual field. After all, the patient said so herself, and there's no reason to think that she was lying. The patient could therefore not possibly know, from her first-person perspective, why her left hand picked up the hammer. We, however, can watch the experiment unfold from our third-person perspective and know exactly what is happening. Science can get inside our heads better than we can! The split-brain experiments also lead to another astonishing conclusion, especially for those who believe in an indivisible, immaterial mind. Split the brain and you'll cut the mind in half, as Sperry and Gazzaniga's experiments so elegantly demonstrate. On the materialistic hypothesis, this conclusion makes perfect sense, since the mind *is* the brain described at a more abstract level.

Split-brain research also provides a unique window onto the kind of confabulation I mentioned earlier. In another classic experiment, a picture of a chicken claw was presented to the "speaking" left hemisphere of a split-brain patient and a picture of a snow scene was shown to his "mute" right hemisphere. The patient then had to manually select from among a set of eight pictures the ones corresponding to what he had seen. Unsurprisingly, his right hand selected the picture of a chicken, because his left hemisphere, the one in charge of the right hand, had seen a picture of a chicken claw. The left hand also chose the appropriate picture, the picture of a shovel, because the right hemisphere had been presented with a snow scene. When asked about his choice of pictures, the patient replied "Oh, that's simple. The chicken claw goes with the chicken, and you need a shovel to clean out the chicken shed."[13]

Let's pause for a second to think about what the patient said. His "speaking" left hemisphere saw the picture of a chicken claw, and so the patient could verbally report on it. He could thus accurately explain why his right hand selected the picture of the chicken. However, since the patient wasn't consciously aware of what was presented to his "mute" right hemisphere—the snow scene—he could not possibly know why his left hand selected the picture of a shovel. But that did not prevent him from offering an explanation for his choice! The explanation is blatantly false but it is nevertheless perfectly consistent with the information that was consciously available to the patient.

The patient consciously knew that he saw the picture of a chicken claw, that his right hand selected the corresponding picture of a chicken, and that his left hand picked the picture of a shovel (the patient could move his head and look at what his hands were doing). Given that information, how could he explain his left hand's choice? By making up a story consistent with the other pieces of information available to his conscious mind. Given what he consciously knew, it made perfect sense for the patient to claim that he picked the shovel to clean out the chicken shed. This kind of confabulation is common with split-brain patients. Present the word *laugh* to a woman's "silent" right hemisphere and she will start laughing. When asked what made her laugh, she'll tell the experimenter that performing tests all day long is a funny way to make a living.[14]

The left hemisphere's propensity to construct explanations for our behavior based on the information available to it, including fictional explanations, is what Michael Gazzaniga has called the "interpreter." Gazzaniga has suggested that the interpreter may be at the origin of our quintessentially subjective feeling of existing as an integrated whole, a self (a soul?), and not just a disparate collection of signals coming from different parts of the brain. Lest you conclude that the confabulatory tendencies of the interpreter can only be revealed in people whose brain has been split, let me tell you about a study done in the early 1960s in which participants (whose brain hemispheres were fully connected) were injected with adrenaline.[15] Adrenaline activates the sympathetic nervous system, resulting in increased heart rate, facial flushing, and shaking of the hands.

The subjects in the experiment interacted with confederates (people who, unbeknownst to the subjects, were in on the experiment) who either behaved as though they were euphoric or angry. The participants who were informed about the effects of adrenaline attributed their symptoms, such as their racing hearts, to the effects of the drug. The subjects who were not informed of the effects of the drug attributed their symptoms to environmental factors. Those who interacted with euphoric confederates reported feeling elated and those who interacted with angry confederates reported being angry. This shows that when an obvious explanation for our feelings and behavior is present, we accept it. But when no such explanation is available, we simply make one up.

Studies like the one I just described demonstrate the role played by the interpreter in people with intact brains. I chose to use the example of split-brain patients to illustrate the role of unconscious mental information, but I didn't really need to rely on such exotic cases, save for the coolness factor. Consider, for example, the sentence you are reading right now. What your eyes are seeing is just ink on a piece of paper or pixels on a screen. Your mind, however, effortlessly interprets the marks it sees and comprehends the thoughts I am trying to convey to you. Could you explain to someone how your mind does that? Unless you are a professional psycholinguist, I seriously doubt that you can. You are aware of the final product, the feeling of understanding what these marks mean, but you are clueless about all the steps your brain has to take in order to transform the images these marks project on your retinas into meaningful thoughts.

This conclusion can be extended to just about any aspect of your mental life. You get the final product, the meaning, the thoughts, the sensations, the perceptions, but you have no idea how your brain performs the trick. Can you explain why your mind insists on telling you that one of Roger Shepard's tabletops is longer than the other? The mind is indeed full of superstition and imposture, and our first-person intuitions are a hopeless guide to its true functioning. In order to understand how the mind really works, we need the third-person perspective offered by science. The failure of our introspective powers to deliver a true understanding of how the mind works is precisely why we need scientific psychology, or cognitive science. If understanding how the mind works was simply a matter of consulting our intuitions, something that everybody can do, psychology departments at universities around the world would close their doors.

The point of this discussion was to show you that we should stop kidding ourselves about the powers of our conscious minds. Just because we all feel very deeply that we are coherent, integrated selves, or souls, that appear to be distinct from the operation of our brains doesn't mean that it is so. Sure, you can trust your intuitions when they tell you that you have found the love of your life. After all, saying that we are in love is merely reporting on the content of our inner mental world. This is not the same as making a claim about how the world works. If every fiber of your being tells you that you are in love, then who in their right mind could tell you otherwise? That would be perverse. However, our conscious mind, when unaided by the powerful instruments and methods of science, cannot tell us how the world really works. Witness the fact that many important scientific conclusions are deeply counterintuitive. Do quarks and gluons, curved space-time, and the double-helix structure of the DNA molecule make much intuitive sense to you? I doubt it. If we really want to know whether people have souls, forget what your parents and your priest told you. It's time to give scientists the microphone.

MAKING UP OUR MINDS

We are indebted to Francis Bacon and his Idols for important insights into the functioning of the human mind. Bacon realized that the mind, unaided by the methods and instruments of science, isn't perfectly calibrated to deliver truths about the objective world. If God gave us a supernatural soul, why should this be so? Why would the mind contain all the imperfections that Bacon warned us about? What Bacon didn't know in the early seventeenth century is that our bodies and brains were shaped not by the hand of God, but by the no-less-invisible hand of natural selection. Natural selection, unlike God, does not care as much about truth as it cares about survival and fitness.[16] What would be the point of having correct intuitions about the behavior of objects approaching the speed of light when everything we care about for our survival moves at ridiculously lower speeds?

Our task in the next chapters will be to determine whether our mental lives arise from the mysterious workings of an immaterial soul or whether they are the product of purely physical processes in our brains. Armed with what we have learned in this chapter, we will be in a better position to answer this question as we proceed with our detective work. First, we will need to recognize that claims about the existence of the soul bear a burden of proof and cannot simply be accepted by default. The Center for Christian Thought at Biola University recently posted a number of videos on YouTube to discuss questions related to the soul. The title of one of these videos, "Does Brain Science Disprove the Soul?" is very revealing. The title presupposes that there is a soul, a conclusion that neuroscience may or may not be able disprove. It is easy to see why the Center for Christian Thought would have a biased opinion on this issue. However, as we discovered in this chapter, the soul hypothesis, a bona fide scientific claim, cannot simply be accepted by default. Like any other factual claim, it has to earn its respectability.

When considering the evidence that has been invoked to support the soul, we will need to ask whether a particular conclusion or line of reasoning counts as decisive evidence for the soul hypothesis. Nothing short of decisive evidence should make us revise our initial skepticism with respect to soul-claims. This is simply the way rational inquiry proceeds in any other sphere of human endeavor. After all, Facilitated Communication was accepted by default and sold to the general public on the basis of bad evidence, with devastating consequences. This is something we will need to bear in mind as we consider the soul hypothesis. Because the soul hypothesis represents a series of claims about physics, biology, psychology, and neuroscience, what should matter in our deliberations are the opinions of professional investigators in those disciplines, what I called scientific Pros. The opinions of scientific Joes, like priests or New Age gurus, do not count.

Then too, when we consider the opinions of professional scientists, we must be careful to differentiate between conclusions that reflect the scientific consensus and those that reflect the opinion of a handful of naysayers.

Finally, we have learned to be wary of our own first-person intuitions. Our conscious minds might insist that we are coherent specks of consciousness distinct from the physical substrate of our brains and bodies, but we have no reason to believe our shallow, shortsighted, and confabulatory minds. In the next several chapters, I will let you examine for yourself the evidence that has convinced the overwhelming majority of scientists in the relevant disciplines that the soul, like phlogiston and the luminiferous ether, is a fiction. As followers of Feynman's first principle, we are now ready to begin examining the evidence.

Chapter 4

DUALISM ON TRIAL

Since our inner experiences consist of reproductions, and combinations of sensory impressions, the concept of a soul without a body seems to me to be empty and devoid of meaning.
—**Albert Einstein, letter of February 5, 1921**

Duncan MacDougall (1866–1920) of Haverhill, Massachusetts, was an early twentieth-century physician who was convinced he had found a way to prove the existence of the soul. MacDougall believed that the soul substance had "gravitative" properties, as he called them, and that the weight of the soul could therefore be determined. The million-dollar question was to find a way to measure such an elusive substance. To fulfill his ambitious goal, MacDougall developed an approach that comes straight out of the X-files of medical science. He reasoned that if the soul leaves the body at the moment of death, it should be possible to determine its mass by weighing a person just before and right after she dies. Amazingly, MacDougall was given permission to perform his macabre experiment on moribund patients suffering from tuberculosis.

A few hours before the patients were expected to die, MacDougall placed their beds onto a gigantic, industrial-strength scale, and he then patiently waited for these poor souls to kick the bucket (I really feel justified in using the phrase here). As the moment of death approached, MacDougall recorded the weight of the patients right before they expired and immediately following the moment of death, after the soul had supposedly left the body. MacDougall performed his test on six patients and recorded a weight loss for four of them. So how much does a human soul weigh? On average, a little less than an ounce, or for those who prefer the metric system, twenty-one grams. As a control, MacDougall carried out the same procedure with nonhuman animals, choosing fifteen healthy dogs as his guinea pigs (he killed

the dogs himself). Dogs, it seems, do not have souls. At least this is the conclusion reached by MacDougall, who observed no weight loss in his canine
sample before and after the moment of death. MacDougall's findings were
published in the April 1907 issue of the medical journal *American Medicine*
with the title "Hypothesis Concerning Soul Substance Together with Experimental Evidence of the Existence of Such Substance."[1]

Today, MacDougall's results are regarded more as an amusing piece of
medical trivia than as decisive evidence for the existence of the soul. This
is because MacDougall's work suffers from a number of serious flaws. For
starters, his sample size was too small (only six human patients), his instruments too imprecise (he had trouble determining the exact moment of
death and could not measure differences in weight precisely enough), and
his results were too variable to allow him to draw any meaningful conclusions. The results from two of his patients had to be discarded and of the
remaining four one showed an immediate drop in weight; two showed that
the initial drop in weight increased over time (how many souls are there in
a human body?); and one showed an immediate drop in weight, followed by
an increase in weight, and then another drop (maybe the soul had trouble
making up its mind). All in all, there was simply too much noise and uncer-
tainty in MacDougall's results to even conclude that the body loses weight at
the moment of death, let alone that the weight loss was due to anything like
the soul leaving the body. (Besides, isn't the soul supposed to be immaterial
and therefore weightless?)

Today, the case for the existence of the soul is made in popular books
and articles written by New Age authors, religious apologists, and spiritual gurus. To be fair, there are also a few naysayers in the sciences who are
unhappy about the existing consensus and claim that materialism is fatally
flawed. Among these recalcitrant minds, we find philosophers, at least one
linguist, and a handful of neuroscientists. This eclectic group of soul advocates has touted a wide range of phenomena to support their dualistic beliefs.
These include: reincarnation and the past lives of young children (very
popular in India); being able to record the voices of deceased individuals by
leaving an active tape recorder on your nightstand (I wonder if this would
work with a smartphone too); the fact that every single cell in our body is

allegedly replaced every seven years but that we nevertheless preserve our identity (hence the conclusion that it must be nonphysical); being able to directly communicate with the dead; near-death experiences; free will and consciousness; people's intuitions; morality; the creative use of language; modern physics; and a plethora of other claims.

In spite of their apparent diversity, soul claims can be grouped into a small number of families that we will examine in this chapter. The beauty of hierarchically organized information (such as family trees) is that once you know that a family of claims is suspect, you can confidently conclude that newly encountered members of the family should likewise be shunned. Thus, if you encounter a soul claim that I have not discussed here, chances are that it will belong to one of the families we are about to examine. In addition to the proposed genealogical approach, I will use a second organizing principle. I will consider families of claims based on phenomena that mainstream science recognizes as real (recall our discussion of Facilitated Communication in chapter 3). The idea is to consider the *strongest possible* cases that have been made for the existence of the soul and show you that even those fall far short of providing decisive evidence. To illustrate, most scientists today do not believe that it is possible to talk to the dead (or at least that such a phenomenon has been demonstrated). Consequently, I will not talk about evidence for the soul based on communication with the dead.

It's not that I am shying away from such business. Rather, what I am trying to do here is little bit like going after the mob. You don't want to spend all your time and resources chasing the low-level thugs. Instead, you want to try to get your hands on the big shots. Once you do, you know that the whole organization will fall apart. In the same spirit, I want to consider the strongest possible evidence for the existence of the soul and show you that even that is utterly unconvincing. In case you are curious about the possibility of talking to the dead, or other paranormal phenomena, there are a number of excellent books written by serious scientists who address these issues. Two classic books that come to mind are Carl Sagan's *The Demon Haunted World* and Victor J. Stenger's *Physics and Psychics*.

With these considerations in mind, there are four families of soul claims that I would like to examine. The first is a family of claims based on introspec-

tion (the self-examination of one's conscious thoughts). As many soul advocates point out, common sense delivers the powerful intuition that our body
and mind are not cut from the same cloth. Does this entail that the mind really
is separate from the body? The second family, a close cousin of the first one,
is based on a fascinating subjective phenomenon called the near-death experience. The phenomenon itself is undoubtedly real. The key question for us is
whether near-death experiences tell us anything about the existence of the
soul. The third family of claims rests on an appeal to recalcitrant phenomena
that elude our current scientific understanding (free will and consciousness,
to name the usual suspects). Since science cannot explain these phenomena,
the argument goes, they must be otherworldly and support the existence of
the immaterial soul. The last family of soul claims we will examine is based on
an attempt to argue that science itself, usually the baffling theories of modern
physics, can be interpreted as lending support to the soul hypothesis.

IN THE "I" OF THE BEHOLDER

Barrington Bayley's short story, *The God Gun*, was in part designed to elicit
the intuition that we are more than mere collections of physical parts. Descartes didn't stretch the boundaries of our imagination when he tried to
demonstrate what feels intuitively obvious to all of us, namely that we are
immaterial minds inhabiting physical bodies. In their book *A Brief History
of the Soul*, philosophers Stewart Goetz and Charles Taliaferro, two of the
New Dualists, remind us that throughout history, the case for the soul has
primarily been made on the basis of what it feels like from the inside to be a
human being. This is what they write:

> In light of our brief history of thought about the soul, which shows that
> the soul's existence is affirmed largely, if not entirely, on the basis of first-
> person experience . . . in other words, given that one is convinced in
> advance (prior to one's consideration of physics) of the truth of dualism,
> it is not the least bit ad hoc or question-begging to search for ways . . . in
> which one might make one's belief in the soul's existence consistent with
> possible developments in physics.[2]

So for Goetz and Taliaferro, common sense convinces us in advance of any kind of scientific investigation that dualism must be true. If it feels true, then it must be true.

In *The Soul Hypothesis*, Stewart Goetz fleshes out this argument by asking us to consider the movement of his fingers on his keyboard as he typed his essay on the soul. Goetz explains that it is obvious that these movements are ultimately caused by his choice to use particular words and phrases for a particular purpose. Since choices and purposes are not physical entities in the world, but rather properties of the immaterial soul, Goetz contends, it follows that what we do, at least voluntarily, is not caused by anything physical, but instead by our immaterial souls. "If our common-sense view of a human being is correct," Goetz explains, "I, as a soul, cause events to occur in the physical world by making a choice to write this essay for a purpose."[3]

But this line of reasoning is utterly fallacious. Just because something feels right, we certainly wouldn't want to conclude that it must therefore be right. Consider, for example, the earth's orbit around the sun. Consult your intuitions and tell me if it feels like the earth is moving around the sun at the speed of sixty-seven thousand miles per hour. It certainly doesn't feel that way. As far as we can intuitively tell, the earth is perfectly immobile. So the proposition that the earth is moving around the sun at a very high speed doesn't agree with common sense, and yet there are very good reasons to believe it is true. Here's another example to illustrate this point. If you have a vivid dream that there is a monster under your bed, also a powerful subjective experience, would you want to conclude that there really is a monster under your bed? The strength of your intuitive conviction, however sincere it may be, is unrelated to the truth of a claim. Comedian Stephen Colbert put it best when he told the crowd at the 2006 White House Correspondents' dinner that "Every night on my show, *The Colbert Report*, I speak straight from the gut, okay? I give people the truth, unfiltered by rational argument."[4]

In chapter 3, I used Roger Shepard's tabletop illusion to show you how unreliable our first-person, subjective impressions can be when compared with the results of objective, third-person measurements. I also gave you reasons to doubt that our conscious mind, unaided by the methods and instruments of science, can deliver reliable conclusions about how the world works. As philosopher Paul Churchland explains, "The red surface of an apple does not look like a matrix of molecules reflecting photons at certain critical wave lengths, but that is what it is. The sound of a flute does not sound like a sinusoidal compression wave train in the atmosphere, but that is what it is. The warmth of the summer air does not feel like the mean kinetic energy of millions of tiny molecules, but that is what it is."[5] Churchland concludes that "if one's pains and hopes and beliefs do not introspectively seem like electrochemical states in a neural network, that may be only because our faculty of introspection, like our other senses, is not sufficiently penetrating to reveal such hidden details."

Goetz's first-person argument for the existence of the soul has it exactly backward. If dualism and what we called the detachability of mind and body is the scientific hypothesis under investigation, we cannot proceed by simply asserting that we are convinced of the conclusion that dualism is true in advance of any serious investigation just because it feels obvious. Just imagine how well this approach would work in a court of law: "Don't bother with all the objective evidence, Your Honor; I am convinced in advance that the suspect must be guilty because every fiber of my being tells me that this must be so." This would be utter madness.

My first objection to arguments for the existence of the soul based on introspection is that this line of reasoning tells us absolutely nothing about the relationship between mind and body. Relying on introspection to deduce that we have souls is falling right into the trap that Francis Bacon tried to warn us about almost four hundred years ago (his Idol of the Tribe). Given what we know today, it would be extremely naive to conclude that just because something feels right it must therefore correspond to a true property of the objective world. In chapter 3, I began to show you that even in the domain of psychology we cannot trust our introspective faculty to deliver accurate insights into the functioning of the mind. Using the example of split-brain patients, and also the more mundane facts of everyday cognition, I showed you that the conscious mind is clueless about what goes on under the hood. Only science, and its reliance on the third-person perspective, can tell us what the mind is and how it really works.

To finish closing the door on the family of soul claims based on first-person arguments, let me now show you why introspection was abandoned by professional psychologists more than one hundred years ago and what came to replace it. In 1879, Wilhelm Wundt, one of the founding fathers of modern psychology, opened the Institute for Experimental Psychology at the University of Leipzig. The goal of Wundt's psychology was to study the structure of the human mind, very much like chemists study the structure of chemical compounds. Wundt believed that conscious mental states could be studied using introspection. After all, people have direct access to the content of their minds. In his research, Wundt asked trained assistants to reflect on the content of their conscious experiences while they were presented with different kinds of stimuli. Wundt insisted that these experiments be conducted under carefully controlled conditions. The participants had to be trained a certain way, and the same stimuli, instructions, and physical surroundings had to be used with different participants.

In an article called "Behaviorism, Methodological and Theoretical," my colleague Randy Gallistel chronicles the demise of introspection and explains why psychology is better off without it. Gallistel explains that Wundt's introspective techniques worked very well until different participants started reporting conflicting impressions when tested under the same experimental conditions. Oops. For example, some participants who were asked to introspect about abstract ideas like honor reported that these ideas are not associated with any sensory content while other participants reported that they are. Given that the aim of Wundt's theories was to identify the fundamental nature of conscious thought, such individual differences were problematic, for they were suggesting that the nature of thought was fundamentally dissimilar in different human minds.[6]

So how was this problem resolved? How would you decide which of two introspecting subjects is right when you get conflicting reports? Unfortunately, you cannot get inside the heads of your participants to find out what is going on. Since the content of a person's consciousness is private and only accessible to that person, there is no way to resolve the dispute. To extricate themselves from this vexing situation, psychologists decided to abandon introspection and rely exclusively on the third-person perspective. Let me explain by revisiting Roger Shepard's tabletop illusion. Recall that when we look at the two tabletops our visual intuitions tell us that one table is longer than the other. This is the conclusion that visual introspection delivers to us.

What are we to do with this introspective verdict? One possibility would be to conclude that our visual intuitions are telling us about the two tables *the way they really are*. If we see one table as being longer than the other, then it must mean that one table *really* is longer than the other. But we know that Shepard's tabletops are an illusion. In reality, the tabletops are identical in length, and our visual intuitions simply fool us into believing that they are not. This is Bacon's Idol of the Tribe at work. Our intuitions also tell us that the earth is immobile, that there is a monster under our bed, and that the mind is immaterial and separate from the body. However, we cannot take any of these conclusions at face value—at least not without corroborating evidence obtained using the third-person perspective of science. This means that we must abandon introspection as a reliable guide to the structure and functioning of the world. If we want to find out how the world works, we need the third-person perspective of science.

Abandoning the idea that introspection can tell us how the world works, however, does not mean that we have to abandon the results of introspection altogether. Indeed, once we understand that Roger Shepard's tabletops are really identical in length, another question immediately arises: Why then do people systematically *perceive* one tabletop as being longer than the other? That's what needs to be explained. So instead of assuming that introspection reveals the world to us as it really is, we need to shift perspectives and ask how we can explain people's subjective reports (however erroneous they may be). This shift in perspectives holds the key to the Wundtian puzzle I described earlier. In order to solve the problem that crippled Wundt's research and put psychology on a more secure scientific footing, its practitioners moved away from the first-person perspective and adopted the third-person perspective. We cannot get inside people's heads, but we can certainly measure their subjective reports and treat them as observable data in need of explanation.

To illustrate, suppose we asked a group of people to look at a picture of Roger Shepard's tables and report on which one they perceive to be longer. People would consult their visual intuitions and, provided that they are not under the influence of alcohol or more exotic chemical substances, they would report that one table is longer than the other. We can then objectively record the fact that all our subjects pointed to the same table as the "longer" one. No need to get inside people's heads. If someone doesn't believe the results of our experiment, she can easily repeat the experiment herself under the same conditions (no drunk subjects and so on) and she too will find that her subjects all point to the same table as the "longer" one. Now instead of regarding the results of people's introspection as providing an explanation of how things really are—in this case, the conclusion that one table really is longer than the other—we treat people's subjective responses *as the data to be explained*. That is, we try to figure out why people reported seeing that one table is longer than the other when we know perfectly well that this isn't true. Again, *that's* what needs to explained.

This distinction between introspection-as-explanation and introspection-as-data-to-be-explained is crucial if we want to avoid falling into the trap of Bacon's Idols and avoid fooling ourselves. Professional psychologists today have completely abandoned the idea that introspection can deliver important truths about how the mind works. The third-person perspective approach to psychology is called *methodological behaviorism*. It is the idea that people's objectively measurable behavior, including reports based on subjective impressions, constitutes the raw data in need of explanation. As cognitive neuroscientist Stanislas Dehaene explains, "As a method, introspection provides a shaky ground for a science of psychology, because no amount of introspection will tell us how the mind works."[7] Dehaene was specifically concerned with the workings of the human mind, but his conclusion applies equally well to the rest of science. Dehaene also points out that "as a measure, introspection still constitutes the perfect, indeed the only, platform on which to build a science of consciousness, because it supplies a crucial half of the equation—namely, how subjects feel about some experience (however wrong they are about the ground truth). To attain a scientific understanding of consciousness, we cognitive neuroscientists 'just' have to determine the other half of the equation: which objective neurobiological events systematically underlie a person's subjective experience."[8]

We now have very good reasons to reject any soul claim based on first-person impressions. Modern psychology, like any other scientific enterprise, operates from the third-person perspective. However, as we just saw, this doesn't mean that scientific psychology is therefore incapable of taking the first-person perspective into account. What I would like to show you now is that the third-person perspective of scientific psychology gives us everything we need to know about the workings of the mind. This should finish convincing you that a person who can inspect the content of his own mind through introspection is in no better position to understand how his mind works than a scientist studying the same mind from the third-person perspective.

To do this, I will adapt an argument given by the philosopher Daniel Dennett in his book *Sweet Dreams*. Dennett invites us to think about the psychological phenomenon of *masked priming*. In a typical masked-priming experiment, participants are presented with a *prime*—for example, a word displayed on a computer monitor for a very brief moment, say forty milliseconds. The prime is sandwiched between two nonsense letter strings, the *masks*, say PgTHwR and TyNJcM, presented immediately before and after the prime. This is followed by the presentation of a target word to which participants have to respond. In one study, Stanislas Dehaene and his collaborators presented participants with numerals like 2, 4, 7, and so on and asked them to respond by pressing a button with one hand if the number was larger than 5 and by pressing another button with their other hand if the number was smaller than 5. What the subjects didn't know was that each target numeral was preceded by a masked prime, which was also a number, and was either larger or smaller than 5.

When the prime and the target were congruent (when they were both smaller or both larger than 5), participants were reliably faster at making their judgment about the target compared to when the prime and the target were incongruent (when one was smaller than 5 and the other larger, or vice versa). Now here's the interesting part: Even when asked to focus on the primes, subjects could not detect their presence or absence, nor could they discriminate them from the strings of nonsense letters. In other words, the primes were perfectly invisible to the subjects' conscious awareness. And yet we know that the subjects unconsciously perceived the primes because they affected their responses to the targets. Dehaene and his team also relied on corroborating electrophysiological and brain-imaging measures to ascertain that unconscious processing of the primes had indeed taken place.

Now let us follow Dennett and think about the implications of phenomena like masked priming for the usefulness of the third-person perspective over the first-person perspective. Consider the set of causes leading from conscious experience to a subject's verbal report of her subjective experience. These are illustrated in (a–c).

(a) Subject's actual conscious experiences

(b) Subject's beliefs about these experiences

(c) Subject's verbal reports expressing those beliefs

Recall from our discussion that the primary data for psychology is (c)—the behavioral responses obtained from subjects. From the subject's first-person perspective, (b) represents the interpreted data (the subject's beliefs about his conscious experience), and this is what needs to be explained. In our masked-priming example, we need to explain why subjects erroneously believed that there were no primes. The distinction between a person's actual conscious experiences, (a), and his beliefs about such experiences, (b), is introduced to account for the possibility that someone may have an actual conscious experience, (a), but may not remember it, and may therefore believe that the experience never happened, (b).

So far, we have seen that the third-person perspective allows us to travel up to (b) in the causal chain. Based on what people tell us, we indeed have access to the beliefs they have about their conscious experiences. Notice, however, that the third-person perspective does not seem to allow us to get all the way up to (a) for the simple reason that we cannot get inside other people's heads to determine whether they really had a conscious experience. At first blush then, it would seem that the third-person perspective is bound to remain incomplete. But a little bit of thinking tells us otherwise. To see this, consider the possible relationships between (a) and (b). The simplest one is that (a) and (b) are always perfectly isomorphic. This means that whenever you really have a conscious experience, (a), you do indeed believe that you did, (b), and vice versa in cases where you do not have a conscious experience. In this scenario, (b) is as good as (a), and so we lose nothing by adopting the third-person perspective.

Suppose now that (a) outruns (b), to use Dennett's terminology. To make things more concrete, imagine that you really did have a conscious experience but that you do not believe that you did because you cannot remember the experience. In this case you would erroneously believe that you never had a conscious experience. If so, your conscious experiences, (a), would be as opaque to you as they are to an external observer. So on this scenario, your first-person perspective gives you no additional information compared to what could be obtained by a third-person observer. Finally, suppose that (b) outruns (a). That is, imagine believing that you really had a conscious experience that you in fact never had. If so, what would need to be explained is (b), your belief, and not your nonexistent conscious experience! So once again, your first-person perspective gives you nothing extra that couldn't be obtained by a third-person observer. All things considered then, there is nothing extra that (a) would give us above and beyond what we could obtain through (b), using the third-person perspective of science. Case closed.

KNOCKING ON DEATH'S DOOR

On October 15, 2012, *Newsweek* proudly announced on its front cover that heaven was real. This startling piece of news was the first in a series of publicity stunts advertising the story of Eben Alexander, a neurosurgeon who claims that his conscious mind was transported into the afterlife while he was in a coma. At the heart of Alexander's story lies a fascinating phenomenon called the near-death experience (NDE), in which people report a range of subjective impressions that are eerily similar to what we imagine the afterlife to be. These unusual experiences involve a feeling that the soul has left the body, an awareness of being dead, moving through a dark tunnel with a world of light at the end, a life review, encounters with deceased people, and a feeling of peace and well-being. Not too surprisingly, NDEs have been interpreted as providing evidence for the existence of the soul and the afterlife. This is the second family of soul claims that we will examine.

The NDE phenomenon has a long history that can be traced back to antiquity. Early reports can be found in the writings of Plato, with an account of a soldier's NDE in *The Republic*. The term NDE itself was coined by the medical student Raymond Moody, whose 1975 book *Life after Life* became a sensational bestseller and brought the phenomenon to the forefront of popular awareness. The publication of Moody's book was followed a few years later by the establishment of the International Association for Near-Death Studies (IANDS) in 1981. This organization promotes the scientific study of NDEs and has its own peer-reviewed publication, the *Journal of Near-Death Studies*. All in all, an estimated eight million Americans claim to have had an NDE,[9] although the actual number may be higher.[10]

So what can NDEs tell us about the existence of the soul? To answer this question, we will need to bear in mind the lessons we learned from the story of Facilitated Communication. As we saw, the existence of typed messages does not, in and of itself, tell us whether these messages came from the autistic individuals or their facilitators. Only under controlled experimental conditions can we settle the question of authorship. This means that we have to separate the existence of a phenomenon from its interpretation. The same logic applies to the NDE phenomenon. Thousands of people across the country and around the world have had unusual subjective experiences. There is no doubt about that. However, as we just established, first-person experiences alone cannot tell us how the mind works and whether it can operate independently from the body. So, as in the case of FC, if evidence based on NDE reports is to count as decisive, it will have to come from controlled experiments.

The New Dualists are quite fond of NDE reports, which they all too eagerly interpret as providing evidence for the existence of the soul. Popular books that purport to "prove" that the soul really exists almost invariably try to dazzle their readers with the mysterious phenomenon of NDE. In *The Spiritual Brain*, for example, Mario Beauregard and Denyse O'Leary offer the following assessment:

> Materialists seem to think that NDEs cannot fit into a materialistic framework, and they must be allowed to be the best judges of that. Nonetheless,

there seems good reasons to believe that mind, consciousness, and self can continue when the brain no longer functions.[11]

Dinesh D'Souza, another champion of the soul and the afterlife, also mentions NDEs in his book *Life after Death: The Evidence*. For him, "on balance, near death experiences do suggest that consciousness can and sometimes does survive death."[12]

The New Dualists are certainly right about one thing: It is not difficult to imagine the kind of evidence that would convince the scientific community that the NDE phenomenon supports the existence of the soul. People who have an NDE often report an out-of-body experience during which they claim that their soul temporarily leaves their body and floats above the room, allowing them to see and hear everything that is happening. These claims make a very clear, testable prediction. If the soul really gets separated from the body during an NDE, it should be possible for the floating self to acquire information that the body itself, from its more restricted vantage point, could not possibly obtain.

Such hypothetical cases would represent veridical instances of NDEs in the sense that the evidence for extraordinary perception could be verified independently through objective, scientific means. This is what is known in the literature on NDE as *apparently nonphysical veridical NDE perception* (AVP). There is anecdotal evidence of such perception. Perhaps the best-known case allegedly involving AVP is the story of Maria and the shoe. The events, which took place in 1977, were reported by Kimberly Clark in 1984 and involve a woman named Maria who suffered two heart attacks. Following the incidents, Maria had an out-of-body-experience during which her "nonphysical self" was allegedly able to spot a tennis shoe on a third-floor window ledge at the hospital where she was being treated. Clark went looking for the shoe, and, amazingly, she was able to find it. Clark added that, given the location of the shoe, it would have been impossible for Maria to spot it from her room unless she pressed her face against the window. (And who would do that while recovering from a second heart attack?)

In 1994, researchers Hayden Ebbern and Sean Mulligan headed to Seattle's Harborview Medical Center where Maria's story took place.

After reviewing the details of Clark's report, interviewing Clark herself, and inspecting the hospital and its surroundings, Ebbern and Mulligan were much less impressed by Maria's story. In 1996, they published a piece in *The Skeptical Inquirer* demonstrating that Clark's report had been embellished and that Maria could easily have spotted the shoe without "leaving" her body.[13] Other spectacular claims have been made about blind individuals who had NDEs during which they could actually see and have veridical perceptions. However, these cases either involve outright fabrication, as in the case of physician Larry Dossey,[14] or they have been found to be no more convincing than the story of Maria.[15] In her 1993 book *Dying to Live,* former believer turned skeptic Susan Blackmore concludes that these cases present no real challenge to a scientific understanding of NDE. Other books, such as anesthesiologist Gerald Woerlee's *Mortal Minds: The Biology of Near Death Experiences* and Mark Fox's *Religion, Spirituality, and the Near-Death Experience*, present detailed arguments demonstrating that NDEs do not provide evidence for the existence of the soul or the afterlife.

In 2009, Janice Holden, Bruce Greyson, and Debbie James published *The Handbook of Near-Death Experiences*. In one of the chapters, "Veridical Perception in Near-Death Experiences," Holden reviews the body of work on veridical NDE, including attempts to establish the reality of the phenomenon under controlled conditions. If you want to determine whether people's souls really leave their bodies during an NDE, one way to find out is to place laptop computers or pictures in locations that only a floating soul could reach, like the top of a closet. When the soul reenters the body, patients should in principle be able to describe what their souls "saw" while floating about in the room. Of the five studies that Holden believes were conducted with the proper controls, she concludes that not a single one of them has been able to demonstrate extraordinary perception of the soul-floating-in-the-room type.

In 2007, Keith Augustine published a series of three comprehensive articles in *The Journal of Near Death Studies* in which he reviewed the evidence from NDEs, including target-identification experiments (what I just described to you), NDEs in the blind, and a host of other cases. Like Blackmore, Augustine concludes that NDEs, however fascinating the phenomenon may be, does not provide any evidence supporting the existence of the soul or the afterlife. In a seemingly never-ending arms race, physician Jeffrey Long and journalist Paul Perry published a 2010 book titled *Evidence of the Afterlife: The Science of Near Death Experiences.* The book is based on a website set up by Long and his wife through which they collected thousands of accounts of NDEs, creating one of the largest databases in the world. According to the authors, NDEs provide compelling evidence for the existence of the afterlife. In his 2012 book *God and the Folly of Faith*, physicist Victor J. Stenger concludes that there is actually no science in Long and Perry's book—just a very long list of anecdotes. As we saw in chapter 3, anecdotal evidence does not count as decisive evidence.

It is fair to say that research on NDEs has failed to convince the scientific community that there is anything paranormal about the phenomenon. There are four excellent reasons to reject the claim that NDEs provide evidence for the existence of the soul and the afterlife. The first, as we just saw, is that there is no body of credible evidence for *apparently nonphysical veridical NDE perception* that would force scientists to conclude that human consciousness can exist and operate independently from the body. The second reason is that there is nothing extraordinary about the experiential manifestations of NDEs. Neuroscientists have shown that the reported subjective experiences associated with NDEs can be linked to well-established neuropsychological processes in the brain.

In a 2011 review article titled "There Is Nothing Paranormal about Near-Death Experiences," published in the journal *Trends in Cognitive Science*, Dean Mobbs and Caroline Watt explain that the frequently reported feeling of being dead is not unique to NDEs but can also be found in a fascinating syndrome called Cotard's Syndrome, or "walking corpse" syndrome, which has been associated with disturbances in the parietal and prefrontal cortex (more on this in chapter 6). In a similar vein, out-of-body experiences (OBEs) are frequently reported by people experiencing sleep paralysis and hypnagogia (vivid dreamlike hallucinations). Moreover, OBEs can be induced by electrically stimulating certain areas of the brain, as the work of Swiss cognitive neuroscientist Olaf Blanke demonstrates.[16] Mobbs and Watt also explain how tunnel vision, meeting with deceased people, and positive emotions have been associated with known pathologies such as Alzheimer's disease and Parkinson's disease, and that there are perfectly plausible neurological explanations for these phenomena.

The third reason to be skeptical of supernatural interpretations of NDEs is that the phenomenon isn't restricted to brushes with impending death. In their article, Mobbs and Watt mention the case of a diabetic patient who, after an episode of hypoglycemia, reported many of the symptoms typically associated with an NDE. Likewise, combat pilots, when subjected to strong acceleration, can experience a phenomenon known as hypotensive syncope, which leads to the kind of tunnel-like vision reported in NDEs. The fourth reason to doubt the paranormal interpretation of NDEs is that there is overwhelming evidence that conscious experience arises from physical activity in the brain, as we will discover in chapter 6. Absent a scientifically credible body of work establishing the reality of veridical NDE perception, there are therefore no grounds for concluding that the unusual experiences accompanying NDEs are anything other than brain-based experiences, however perplexing they may be.

Returning to Eben Alexander, what can we say about his alleged voyage to heaven? Is he the new Galileo of neuroscience or just the latest fad in a long series of failed attempts to establish the existence of the soul based on NDE reports? The latter, I am afraid. I have read his *Newsweek* article and the book he wrote about his experience, *Proof of Heaven*. If you are looking for a modern-day fairytale, then you should read the book too. It is well written, entertaining, and a quick read. If you are scientifically curious, however, brace yourself for a major letdown. There is nothing in Alexander's book that even remotely resembles proof. Alexander's fantastical claims rest on three premises: He had a very vivid subjective experience (first premise); he claims that his brain had completely "shut down" while that happened (second premise); and he is a neurosurgeon, so we're supposed to find his story credible (third premise).

Alexander's first-person experience and his status as a neurosurgeon tell us absolutely nothing. Just think about how little weight these two factors would have in a court of law. "The suspect must be guilty, Your Honor." And after the judge asked how you knew that: "Well, I had a dream about it. Trust me, I'm a neurosurgeon." As for the claim that his brain had completely "shut down" during his experience—inviting the conclusion that his NDE wasn't brain-based—it is equally implausible. What about the rich memories Alexander describes in his book? Where did they come from? The minute you say that Alexander's memories are brain-based, his entire account falls apart—which is why he insists that his brain had completely shut down and that his experience cannot have been brain-based.

Even if we were to grant Alexander the supremely implausible conclusion that his brain had completely shut down during his coma, he has absolutely no way of knowing that his whole experience didn't happen as his brain was shutting down or while it was coming back online as he regained consciousness. At the end of his book, Alexander considers this scenario but quickly dismisses it by claiming that "given the intricacies of my elaborate reflections, this seems most unlikely."[17] In other words, Alexander offers no argument and simply asks us to take his word for it—hardly the approach that a serious scientist would take. Sam Harris, one of the New Atheists, was asked whether he would debate Eben Alexander on the question of NDEs and the afterlife. After pointing out the glaring flaws we just reviewed, Harris responded that there was nothing to debate.[18]

THE SOUL OF THE GAPS

The third family of soul claims we will consider involves a fallacy that has been described as the god-of-the-gaps argument, the argument from ignorance, or the argument from personal incredulity. Before showing you how these arguments are used by the New Dualists, let me give you a feel for their general form. Thanks to Fox News celebrity Bill O'Reilly, the logic of god-of-the-gaps argument has become viral. During an interview with David Silverman, president of the American Atheists, O'Reilly challenged his baffled guest to explain how the tides so predictably and regularly go about their business. "You can't explain that!" O'Reilly told Silverman.[19] But if you assume that God exists, as O'Reilly says he does, then everything makes perfect sense and you can understand how the tides work.

Some people (known as "pinheads" in the anchor's colorful jargon) informed O'Reilly that we *do* know how the tides work, and that we have known for more than three hundred years. Undaunted, O'Reilly posted a clip on YouTube in which he pushed the argument one step further.[20] Fine, he concedes, the gravitational pull of the moon explains the tides. But where did the moon come from? (For an amusing parody demonstrating the existence of mail fairies using O'Reilly logic, YouTube is also an excellent resource—because you can't explain the superb regularity with which the mail gets delivered, and if you can, you sure can't explain where the mailman came from.)

The fallacy then, is to automatically assume that if something challenges our current scientific understanding, or the limited grasp of science on the part of the person making the claim, a supernatural explanation wins by default. As the physicist Jean Bricmont explains, the god-of-the-gaps argument merely amounts to giving our ignorance a name—God, the soul, ancient aliens, and so on (recall our discussion of the perimeter of ignorance in chapter 2). Bricmont points out that dogs do not understand the laws of celestial mechanics, but this does not mean that anything supernatural must be involved.[21] Instead, if you want to make a case for God, the soul, or ancient aliens (who supposedly helped the Egyptians build the pyramids), you need to provide positive evidence for the existence of these ideas in accordance with the logic of the burden of proof. Merely pointing to "gaps" in our scientific understanding accomplishes nothing.

Let me now show you how the New Dualists rely on the flawed logic I just described to try to make their case for the existence of the soul. We'll consider two attempts to use ignorance to sell soul beliefs, one made by a scientific Joe, Dinesh D'Souza, author of *Life after Death: The Evidence*, and the other by a scientific Pro, Mark Baker, coeditor of *The Soul Hypothesis*. We shouldn't be too surprised to find that scientific Joes rely on flawed logic to make their case. After all, these people are not trained scientists, and so they can be forgiven for not fully understanding the rules of the game. What is far more revealing is to see that even professional scientists can do no better than offer fallacious arguments to support their soul claims.

So let's meet our first contender, D'Souza's argument from "cosmic justice." D'Souza begins by identifying what he believes to be a gap in our scientific understanding of human nature. This lacuna, D'Souza believes, is morality. According to him, "Evolutionary theories . . . utterly fail to capture this uniquely human sense of morality as duty or obligation."[22] Before presenting his case, however, D'Souza reassures his readers that he will not be relying on the flawed logic of the god-of-the-gaps. "Skeptics at this point may scorn my claim that certain features of human nature seem to defy scientific explanation. They will claim that I am making a discredited appeal to the "God of the gaps."[23] Instead, D'Souza explains that he will rely on what he calls a presuppositional argument. Now what is that? According to D'Souza, presuppositional arguments are what scientists invoke to solve the problems they face. For D'Souza, presuppositional arguments are bold new hypotheses that explain the relevant facts better than any alternative hypothesis.

Copernicus and Einstein used presuppositional arguments to fill "gaps" in our understanding of the world, D'Souza tells us. So if this approach worked for some of the greatest scientific geniuses, then surely we cannot fault D'Souza for using it as well. So what is D'Souza's bold new hypothesis? Here it is: "My hypothesis on offer is, 'There has to be cosmic justice in a world beyond the world in order to make sense of the observed facts about human morality.'"[24] According to D'Souza, human beings, believers and atheists alike (phew!), inhabit two worlds. Realm A is the evolutionary world and realm B is "the other world," the world of comic justice, that D'Souza believes is "built into our natures"[25] (obviously not through any known genetic mechanism).

And what exactly does D'Souza's bold new hypothesis explain? Well, morality of course! For example, the cosmic realm that D'Souza invokes "makes us dissatisfied with our selfish natures and continually hopeful that we can rise above them."[26] The existence of "cosmic justice" also helps us understand why people so often violate morality. "The reason," D'Souza explains, "is that our interests in this world are right in front of us, while the consequences of our actions seem so remote, so distant, and thus so forgettable."[27] I wish there were more to say, but that's about it. D'Souza, who claims in his book that his case is "entirely based on reasoned argument and mainstream scholarship,"[28] doesn't seem to understand the most elementary aspects of the scientific enterprise, such as what constitutes a valid hypothesis, how it bears on the relevant set of facts, and what a valid explanation looks like. This may explain why D'Souza calls his approach "Christian martial arts," fancying himself a "Christian cage fighter."[29]

Unlike a "Christian cage fighter," a scientist would immediately recognize that D'Souza's hypothesis is so vague that it is compatible with almost any set of facts. Take, for example, his claim that our interests in this world are right in front of us whereas the consequences of our actions are so remote (presumably because we have to die first, before fully entering the world of cosmic justice). This, according to D'Souza, is supposed to "explain" why people so often violate morality. But why don't people violate morality *all the time* then? If people did, this would be equally compatible with the idea that cosmic justice and its consequences are "so remote, so distant, and thus so forgettable." This is where the comparison with Copernicus and Einstein might have been a tad exaggerated. As for the set of relevant facts about morality, there is an entire scientific literature on moral cognition full of experimental results obtained with regular folks, children and babies, and people who suffer from various pathologies. Those are the facts that any serious hypothesis would try to explain. Instead, D'Souza serves us nothing but vague platitudes like being "dissatisfied with our selfish natures" and being "continually hopeful that we can rise above them." If we think of the scientific literature on moral cognition as the iceberg representing morality, D'Souza isn't even close to touching the tip; he's in the airplane flying thirty-five thousand feet over the surface of the ocean.

Here's my own version of the presuppositional argument. Forget about D'Souza's world of cosmic justice. Morality is mysterious indeed, but if you entertain the bold new hypothesis that our decisions are influenced by a little angel sitting on our right shoulder and a little demon sitting on our left shoulder, then everything makes perfect sense. Why do people sometimes violate morality? D'Souza asks. Because they listen to the demon instead of the angel. And why do people sometimes act morally? Because they listen to the angel instead of the demon. In fact, my hypothesis is superior to D'Souza's because it can even explain why people sometimes experience difficulty making the right moral choice. This is because they are unsure whether they should listen to the angel or to the demon. What I just offered is a caricature, of course, but my point is that D'Souza's argument for comic justice is no better. Finally, notice that D'Souza's argument, his third "proof" for the existence of the soul and the afterlife (we'll consider his argument from neuroscience in chapter 6), doesn't tell us anything about the soul. Even if we granted him the existence of his world beyond our world, this would tell us absolutely nothing about whether we have a detachable soul that is actually capable of traveling to that world when we die.

Let us leave the world of Christian martial arts to see how scientific Pros use the soul-of-the-gaps argument. I have chosen to examine an argument made by linguist Mark Baker in *The Soul Hypothesis*, a book that he coedited with fellow dualist Stewart Goetz. In a chapter of that book, called "Brains and Souls; Grammar and Speaking," Baker presents an argument for the existence of the soul based not on morality but on our ability to use language creatively. Baker starts with the assumption that the existence of the soul should be regarded as an open question:

> Suppose, then, that we start from the position that whether or not people have souls could go either way. It is something like a 50/50 bet, with neither view on the matter having an especially high burden of proof to meet with respect to the other. Beginning from this neutral position, it is natural and appropriate to treat the question about souls as the Soul Hypothesis—as something that could play a role in guiding scientific research and other forms or rational investigation.[30]

Baker proposes that the soul hypothesis can be used as a guiding principle encouraging investigators to look for mental phenomena that do not (entirely) depend on brains and bodies. To the extent that such phenomena can be identified, they would in turn lend credence to the hypothesis that led to their discovery in the first place—the soul hypothesis. Alternatively, failure to find such phenomena may be interpreted as lending support to the materialistic thesis.

As you can imagine, Baker claims to have identified a phenomenon of exactly the right type. (Otherwise why would he have bothered writing his book chapter?) The phenomenon that Baker has in mind is called the *cre-*

ative aspect of language use, or CALU. Here's how professional linguists think about it. In their work on the human language faculty, linguists identify three main components: the lexicon (our mental dictionary), the grammar (a set of rules for combining elements from the lexicon), and the CALU. Baker illustrates these notions with an analogy from the construction industry. He asks us to think about the lexicon as the bricks and mortar needed for our construction projects. The grammar would be akin to the building codes and engineering principles necessary to build larger structures such as walls, roofs, and houses. Finally, the construction industry also needs architects and contractors to decide where the walls go, how big the rooms should be, and how many bathrooms to build. The creative freedom of the architects to design structures that are consistent with but not determined by the building codes and engineering principles is what Baker equates with the CALU in his construction analogy.

To illustrate the operation of the CALU without making the discussion too technical, let me use a simple example. To begin, think of a response that is fully determined and predicted by the stimulus that triggered it. Let's say someone shines a light in your eyes and your pupils contract. Or your doctor uses a reflex hammer to check your nervous system and the portion of your leg below your knee predictably jerks up. Now think about the amazing freedom and flexibility that language affords us compared to the simple reflexive behaviors I just described. Suppose that one of your friends picked up ceramic painting as a new hobby. Upon being presented with her latest creation (the stimulus), one would be hard-pressed to predict exactly what you would say (the response). Indeed, you may be polite and reply "Very nice piece!" or sarcastic and say "How talented you are!" or indifferent and tell your friend "Great, let's have dinner now." While our use of language is not caused by the relevant stimulus (in any predictable way), it is not completely random either. If it were, it would simply be impossible to communicate. Instead, our use of language is *appropriate* to situations (notwithstanding the occasional slips, of course).

The next step in Baker's argument is to consider standard materialistic evidence that would suggest that a particular mental capacity is physically realized in the brain. To do so, Baker examines the neuroscientific, genetic and evolutionary evidence bearing on the CALU. Reviewing evidence from aphasias (impairments of the language ability), genetically-based developmental disorders, and the "language" of the apes, Baker concludes that the CALU seems to have all the properties that it should have if it is indeed soul-based. That is, Baker points out that there does not seem to be any brain circuitry dedicated to the CALU that could be differentially affected by brain injury or perturbed by genetically based developmental disorders. Moreover, even the smartest of the apes do not seem to possess CALU. Baker concludes that "it is thus reasonable to think that the CALU will fall out as the interaction between soul theory and the theory of language."[31]

I chose to review Baker's argument to show you that even professional scientists working on the human mind can do no better than offer soul-of-the-gaps arguments. This is because there is no positive evidence that would force a rational person to conclude that souls exist. And so instead, reveling in our collective ignorance, the more sophisticated New Dualists point to recalcitrant phenomena that science hasn't yet elucidated. (Remember our discussion of the perimeter of ignorance in chapter 2?) But Baker himself knows perfectly well that his CALU-based argument does not count as decisive evidence for the existence of the soul. As he writes at the end of his chapter, "There is thus no conclusive proof that we have souls distinct from brains to be found in this material."[32]

Curiously, Baker adds that this is nevertheless "very different from saying that we now know beyond a reasonable doubt that the brain is everything there is."[33] This may or may not be what other scientists claim, but this is largely irrelevant. In the end, Baker's worries about exaggerated claims regarding what is or isn't brain-based are dwarfed by his own casual assumption that the existence of the soul is like a "50/50 bet, with neither view on the matter having an especially high burden of proof to meet with respect to the other." As I will show you in the next two chapters, you would need to bury your head in the sand pretty deeply to conclude that the soul hypothesis is a 50/50 bet.

BEND IT LIKE BECKHAM

The last family of soul claims we will examine is nicely captured by a quote attributed to Abraham Lincoln. The sixteenth president of the United States remarked that you can fool some of the people all of the time, and all of the people some of the time. However, Lincoln added, "you cannot fool all of the people all of the time"—thank goodness! As we saw in the previous chapter, science is by far the best way we have developed to avoid fooling ourselves about the workings of nature. In principle, the general rules of the scientific game are straightforward. To be taken seriously, proponents of new ideas have an intellectual obligation to convince their scientific peers of the validity of their ideas. If they cannot do that, they simply have no case. Advocates of Facilitated Communication can insist until they are blue in the face that their technique is valid, but unless they can demonstrate that it works under rigorously controlled experimental conditions, they have no bragging rights.

In practice however, there are loopholes that merchants of superstition can exploit to sell their extraordinary claims. The most common approach is to take advantage of the distinction between scientific Pros and scientific Joes that we explored in chapter 3. Since most people aren't professional scientists, fooling them is much easier than fooling trained professionals. (Again, I am not saying that regular folks are deficient in any way. All I am saying is that most people aren't trained scientists.) There are two specific implementations of this general strategy that we will examine. The first consists of blurring the line between mainstream science and pseudoscience.

In chapter 3, I warned you about scientific naysayers like Daryl Bem. Most scientists can see right through Bem's flawed research, and virtually everyone in the sciences dismisses his extraordinary conclusions. But you can see how easy it would be to convince the general public that Bem's conclusions should be taken seriously. After all, Bem is a real psychologist, with real academic credentials, and he even managed to publish his work in a real peer-reviewed journal. So the first strategy is this: make a list of your favorite Daryl Bems and cite their work as conclusive evidence that your extraordinary claims are backed up by "science." It also helps to portray your heroes as unrecognized Galileos and decry the close-minded and dogmatic nature of mainstream science in its refusal to recognize the work of your champions as legitimate.

Let us take a look at how the New Dualists play this game. In *Proof of Heaven*, Eben Alexander explains to his readers that mainstream scientists are ignorant of all the evidence supporting his own dualistic conclusion. He writes that "those who assert that there is no evidence for phenomena indicative of extended consciousness, in spite of overwhelming evidence to the contrary, are willfully ignorant."[34] First red flag: the predictable charge that mainstream science is ignorant and refuses to accept the "overwhelming evidence" supporting Alexander's favorite conclusion. So where can the eager reader find the relevant evidence? According to Alexander, the truth is presented in a 2007 book called *Irreducible Mind: Toward a Psychology for the 21st Century*, a book that he praises and describes as containing "rigorous scientific analysis."[35] Second red flag: *Irreducible Mind* is a lengthy repository of phenomena that mainstream science either does not recognize as legitimate or does not believe involve anything paranormal. The phenomena mentioned in the book include reincarnation, stigmata, sudden graying of the hair, maternal impressions, near-death experiences, out-of-body experiences, apparitions, and of course, *psi*.

New Dualists Mario Beauregard and Denyse O'Leary follow the same party line in their book *The Spiritual Brain: A Neuroscientist's Case for the Existence of the Soul*. Their argument begins with the usual refrain that mainstream science is dogmatically narrow-minded and that "data that defy materialism are simply ignored by many scientists."[36] And again we hear about the "evidence" for *psi* that mainstream science and its materialistic ideology refuses to take seriously. Yes, mainstream science rightfully ignores extraordinary claims about *psi* or Facilitated Communication, for all the reasons that I explained in chapter 3. Some of the New Dualists would have us believe that there is "overwhelming evidence" supporting *psi*, but this conclusion is pure fantasy. We do know what it means to have "overwhelming evidence" for something. Take electromagnetism for example. The reality of the phenomenon is verified every day by millions of people who use cellular phones worldwide, and there are mathematically precise theories that describe the behavior of electromagnetic waves. *That's* overwhelming evidence.

If *psi* were real, then why don't we have comparable evidence? Why don't people who can allegedly "see" the future bankrupt all the casinos on the planet or use their skills to beat the stock market? The reason this never happens is that nobody can "see" the future, because precognition is ruled out by everything we know about modern physics, as we saw in chapter 3. So yes, mainstream science ignores extraordinary claims made by people like Daryl Bem, not only because the research itself is flawed but also because the conclusions, given what we know about modern science, are about as plausible as the claim that a well-trained athlete can run a marathon in less than ten minutes.

Let us now examine the second strategy that the New Dualists rely on to sell their extraordinary claims. This is what I call the metaphysical game. Here's how it works: Any scientific theory can be used as a set of tools to make predictions about events in the natural world, but it can also be "interpreted" as telling us something about the nature of reality (rightly or wrongly, of course). I touched on this in chapter 2 when I described the difference between methodological and metaphysical naturalism. When a scientific theory is particularly complex and mysterious, merchants of pseudoscience can exploit the resulting weirdness to their advantage and interpret the theory as providing support for their own extraordinary conclusions. No need to rely on controversial phenomena like *psi* or NDEs. All you need is a perfectly legitimate scientific theory to which you can give a metaphysical spin that supports your own biases. As before, it is important to pitch your conclusions directly to the general public, and it is even better to describe your ideas using impressive-sounding terms borrowed from the theory itself.

Enter quantum mechanics, arguably the most spectacular scientific theory ever developed. Quantum mechanics is the branch of physics that deals with the microscopic world of atomic and subatomic phenomena. In addition to making amazingly accurate predictions, quantum mechanics is also notoriously difficult to understand, and it is deeply counterintuitive. As the great Richard Feynman once said "I think I can safely say that nobody understands quantum mechanics."[37] For merchants of superstition, quantum mechanics is the gift that keeps on giving. Because it is an established scientific theory, they can mention it with a straight face and claim scientific legitimacy. And because it is so mysterious, they can point to its counterintuitive conclusions and confidently assert that they support their own extraordinary conclusions. The approach reminds me of an old commercial for ginger ale that goes something like, "It looks like alcohol, it tastes like alcohol, but it isn't." Replace the word *alcohol* with the word *science*, and you get the idea.

Physicist and Nobel laureate Murray Gell-Mann coined the term "quantum flapdoodle" to describe abuses of quantum mechanics to support extraordinary claims. In his long battle with pseudoscience and superstition, physicist Victor J. Stenger wrote several books showing that quantum mechanics does not prove the existence of God, paranormal phenomena, or the independence of mind and body. These books include *Physics and Psychics*, *The Unconscious Quantum*, and *Quantum Gods*. I've looked at five prominent books by New Dualists in which the authors claim to demonstrate the existence of the soul. These are the four books I already mentioned, by Dinesh D'Souza, Mark Baker and Stewart Goetz, Mario Beauregard and Denyse O'Leary, and Eben Alexander, and a 2006 book by New Age guru Deepak Chopra called *Life after Death: The Burden of Proof*. Without fail, all of these books mention quantum mechanics to either directly or indirectly suggest that modern physics proves the existence of the soul.

In his book, Dinesh D'Souza, who is not a physicist, confidently asserts that "far from undermining the chances of life after death, modern physics undermines the premises of materialism."[38] This may be so in D'Souza's own universe, but in the real world, physicist Victor J. Stenger tells us that "Modern physics, including quantum mechanics, remains completely materialistic and reductionistic while being consistent with all scientific observations."[39] My favorite example of a New Dualist caught in the act of peddling quantum flapdoodle involves New Age guru Deepak Chopra. Chopra took part in a panel discussion in which he tried to explain to the audience how quantum mechanics supports his extraordinary claims about the nature of mind and reality. Unknown to Chopra, the theoretical physicist Leonard Mlodinow was in the audience. During the Q&A session, Mlodinow stood up and politely asked Chopra if he would be interested in taking a short course on quantum mechanics to straighten out his interpretation of the theory.[40]

Fortunately for the nonspecialists among us there is an easy way to decide whether we should believe the New Dualists when they assure us that quantum mechanics demonstrates the reality of dualism. Since quantum physicists do not perform experiments on *psi*, near-death experiences, or communication with the dead, which could in principle settle the question of mind-body detachability, the New Dualists instead rely on the metaphysical interpretation of quantum mechanics to make their case. By and large, the New Dualists are scientific Joes when it comes to quantum mechanics, and so all you need to do in order to assess the credibility of their claims is figure out how actual quantum physicists, the scientific Pros, interpret quantum mechanics. If the opinions and conclusions of legitimate experts in the field of quantum physics match the confident assertions made by the New Dualists, then we have something to talk about. If not, you might as well ask your cat what she thinks about quantum mechanics.

A recent article published in the journal *Annalen der Physik*, one of the oldest physics journals, sheds light on this question. The authors report the results of a poll they performed among experts on quantum mechanics during a 2011 conference on the topic of quantum physics and the nature of reality. Here's how the authors summarize their main finding: "Quantum theory is a phenomenally successful cornerstone of physics. But a recent poll among quantum physicists shows that there is still little agreement on what the theory tells us about physical reality."[41] Later in the article, they write that "the fundamental questions, it seems, still elicit radically different answers."[42] This doesn't mean that physicists disagree on how to use quantum mechanics; they only disagree about the metaphysical interpretation of the theory. If there is little agreement even among quantum physicists about what quantum mechanics tells us about the nature of reality, how can a small group of nonphysicists, the New Dualists, possibly be so confident about their own interpretation of a theory that most of them do not even understand? Fortunately, you do not need to be a quantum physicist to answer this question.

Duncan MacDougall may have been one of the first modern scientists to try to establish the existence of the soul experimentally. But he failed. And so has everybody else. Today, there are no scientifically credible phenomena that demonstrate the detachability of body and mind. The New Dualists are right in thinking that the soul hypothesis is a scientific one. But they are seriously fooling themselves when they conclude that there is evidence supporting their hypothesis. There isn't. And so to make their case, the New Dualists have no other choice than to bend mainstream science beyond recognition, rely on the first-person fallacy, the flawed evidence from NDEs, or the warped logic of the soul-of-the-gaps. In the next chapter, I will show you that, far from lending support to the soul hypothesis, the discoveries of modern science actually fly in the face of dualism. To drive the final nail in the soul's coffin, I will show you in chapter 6 that the materialistic hypothesis we defined in chapter 3 is supported by an overwhelming amount of converging evidence.

Chapter 5

REQUIEM FOR THE SOUL

There are many hypotheses in science which are wrong. That's perfectly all right; they're the aperture to finding out what's right.
—**Carl Sagan**, *Cosmos: A Personal Voyage*, 1980

In *Essays upon Some Controverted Questions*, Thomas Henry Huxley, a nineteenth-century anatomist nicknamed "Darwin's bulldog" for his passionate defense of evolution, offers the following advice: "Trust a witness in all matters in which neither his self-interest, his passions, his prejudices, nor the love of the marvelous is strongly concerned. When they are involved, require corroborative evidence in exact proportion to the contravention of probability by the thing testified."[1] A century before Huxley, the Scottish philosopher David Hume expressed the same idea in these words: "A wise man proportions his beliefs to the evidence."[2] Today, the idea is captured by Carl Sagan's catchy slogan that extraordinary claims require extraordinary evidence. As we saw in the previous chapter, one of the reasons mainstream science has abandoned the soul is that the claim is extraordinary but the evidence is not.

However, this is not the only reason for the demise of the soul. There are two additional, mutually reinforcing lines of reasoning that have led mainstream scientists away from the soul. The first is anchored in what modern science has taught us about the world. An idea, old or new, may not be supported by the kind of evidence that would make scientists accept it, but it may nevertheless be compatible with what is known about the natural world, as revealed to us by science. In his book *Physics of the Impossible*, theoretical physicist Michio Kaku sees such ideas as belonging to class I impossibilities. These ideas may be far-fetched, but they do not violate the fundamental laws of physics, and so they may become a reality in the future. Kaku believes that teleportation, antimatter engines, and invisibility are class I impossibilities.

At the other end of the spectrum, we find class III impossibilities. These correspond to ideas that *do* violate the laws of physics, and so discovering that such ideas are indeed true would entail a major overhaul of the foundations of science. For Kaku, perpetual-motion machines and precognition count as class III impossibilities.

According to the dualistic hypothesis, the soul is not only detachable from the body, which allows it to survive after we die, but it is also claimed to be nonphysical (or immaterial). We saw in chapter 4 that there is no scientifically credible evidence for the detachability of body and mind. What I would like to show you in this chapter is that the notion of an immaterial substance that can causally interact with the body flies in the face of what we know about modern science. Viewed in terms of Kaku's continuum, souls come tantalizingly close to falling in the category of class III impossibilities. Worse, the concept of an immaterial soul substance has no useful formulation, if it even has a coherent one, and it is therefore utterly devoid of any explanatory power. Together, these conclusions represent the second line of reasoning that has led mainstream science to abandon the soul. In the next chapter, we will drive the final nail in the soul's coffin and explore the wealth of evidence supporting the materialistic hypothesis—dualism's nemesis. But before we get there, let us consider in more detail how the idea of a causally potent, immaterial substance meshes with what we have learned from modern science. To do so, we'll begin with Thomas Huxley's favorite topic, the world of evolutionary biology.

EVOLVING SOUL

In the seventeenth century, when Descartes formulated his version of dualism, it was still possible to believe that only human beings have souls. This is because Descartes's contemporaries also believed that human beings were created separately from the rest of the animals. This conclusion was buttressed by the observation that we seem to possess faculties that even the most intelligent of the beasts lack. Language, to use an example we touched on earlier, is such a capacity, as Descartes already pointed out. What Descartes could not have predicted, however, is that two centuries later creationism received its *coup de grâce* with the publication of Darwin's ideas. Since then, the conclusion that species are not immutable has been elevated to the status of a fact, and the evidence that human beings evolved from more modest life-forms is overwhelming, to put it mildly.

Evolution then creates a problem for the idea of the soul. As the philosopher Paul Churchland explains, "The standard evolutionary story is that the human species and all of its features are the wholly physical outcome of a purely physical process."[3] Given this account of our origins, Churchland continues, "There seems neither need, nor room, to fit any nonphysical substances or properties into our theoretical account of ourselves. We are creatures of matter."

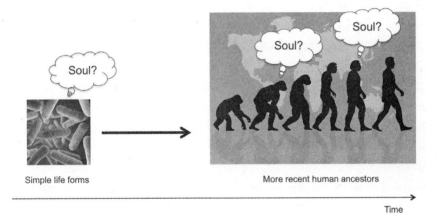

Figure 5.1: Evolution and the Soul.

Now that the seemingly impenetrable wall that was once believed to separate us from the rest of the biological world has fallen, the claim that we contain an additional, immaterial ingredient becomes supremely implausible. At what point in the unbroken chain between primitive life-forms and human beings did the soul get added to the mix and why? (See figure 5.1.) To make things worse for dualism, the divide between the world of inert matter and the world of living organisms, which compelled Aristotle to postulate his life souls, has also been bridged. Writing during the first half of the twentieth century, Bertrand Russell explained that the mechanistic conception of life was already sufficiently accepted by the majority of working scientists to be the view described in the *Encyclopedia Britannica*. Russell cites the following entry for evolution: "A living organism, then, from the point of view of the scientific observer, is a self-regulating, self-repairing, physicochemical complex mechanism. What, from this point of view, we call 'life' is the sum of its physicochemical processes, forming a continuous interdependent series without break, and without the interference of any mysterious extraneous force."[4] The physical basis of life, coupled with the theory of evolution, poses a formidable challenge to dualism. We are the physical outcome of a purely physical process operating on the basis of a purely physical set of raw ingredients. There is no need or room for any nonphysical substances in this equation.

THE INTERACTION PROBLEM: UNINTELLIGIBILITY

Let us leave the world of biology to begin thinking about what the physical sciences have to say about the soul. After reading *Meditations on First Philosophy*, Elizabeth of Bohemia, the eldest daughter of Frederick V, corresponded with Descartes for several years on matters of common intellectual interest. In a letter written in May 1643, the astute princess achieved philosophical fame by putting her royal finger on a particularly vexing problem for dualism:

> I beseech you tell me how the soul of man (since it is but a thinking substance) can determine the spirits of the body to produce voluntary actions. For it seems every determination of movement happens from an impulsion of the thing moved, according to the manner in which it is pushed by that which moves it, or else, depends on the qualification and figure of the superficies of this latter. Contact is required for the first two conditions, and extension for the third. You entirely exclude extension from your notion of the soul, and contact seems to me incompatible with an immaterial thing.[5]

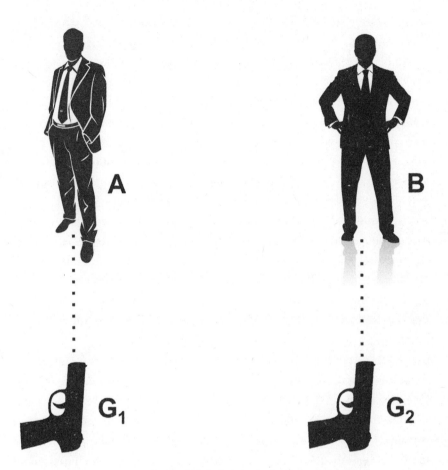

Figure 5.2: Bullets in Action.

Recast in more modern parlance, the princess asked how the immaterial soul substance could possibly interact with physical matter, as Descartes believed it did (recall that Descartes had even located the seat of such interaction in the pineal gland). Think of Patrick Swayze in the movie *Ghost*. He faced the same problem when he tried to move physical objects around him. Try as he might, his ghostly foot would go right through the can he was trying to kick. Our friendly ghost Patrick fared no better when he tried to push a quarter with his translucent finger. In the movie, Swayze eventually meets another ghost who teaches him how to move physical objects by "concentrating" really hard. Since we live in the real word and can't rely on Hollywood magic, we will need to take the interaction problem seriously.

In his book *Physicalism or Something Near Enough*, philosopher Jaegwon Kim puts some flesh on the interaction problem's bones. Kim begins by observing that among philosophers, the interaction problem is widely recognized as dualism's Achilles' heel—an insurmountable problem that single-handedly led to the demise of the Cartesian doctrine. To give the problem some teeth, Kim relies on an ingenious thought experiment in which he asks us to imagine two guns, G1 and G2, being fired simultaneously and resulting in the death of two individuals, Adam and Brian (see figure 5.2).

Returning to my *CSI: Miami* analogy, imagine that we were interested in determining which gun killed Adam and which gun killed Brian. How could we settle this question? One approach would be to trace a continuous causal path between G1 and Adam and between G2 and Brian. This way, we could unambiguously determine that the bullet fired from G1 killed Adam and that the one fired from G2 killed Brian. This could be accomplished by using a high-speed camera to follow the trajectory of the bullets in slow motion. By tracing a continuous causal path between each bullet and its target, we could establish beyond the shadow of a doubt that G1 killed Adam and that G2 killed Brian.

But suppose we didn't have a high-speed camera and we couldn't retrace the trajectory of the bullets in slow motion. How else could we determine that G1 killed Adam and that G2 killed Brian? Given what we know about guns and ballistics, we could reason that the spatial orientation of the guns and their distance relative to their targets would be sufficient to establish that G1 killed Adam and that G2 killed Brian. To see this, look at figure 5.2 again. It follows from the basic physics of guns that a bullet fired from G1 could not possibly have killed Brian, barring a "magic bullet" scenario or a spectacularly improbable series of ricochets. The same holds for G2 and Adam. Thus, given the relative position of the guns and the two individuals, we would fully expect a bullet fired from G1 to hit Adam and a bullet fired from G2 to hit Brian. Couched in slightly more abstract terms, the reasoning we just engaged in amounts to finding a pairing relationship that holds between G1 and Adam and G2 and Brian, but not between G1 and Brian and G2 and Adam. Notice that our pairing relationship is defined in terms of the framework of space. G1 is correctly paired with Adam—but not with Brian—because it is properly *angled* toward Adam and it is also within the right *distance* of Adam (and likewise for G2 and Brian).

Next Kim asks us to consider a situation in which two souls, X and Y, cause two individuals, Charles and Denis, to voluntarily move their arm through an act of free will (see figure 5.3). In case you are wondering why you cannot see the souls in figure 5.3, let me reassure you that they are there and tell you that you are asking exactly the right question. Bear with me for another paragraph or two, and everything, like the souls in figure 5.3, will become transparently clear.

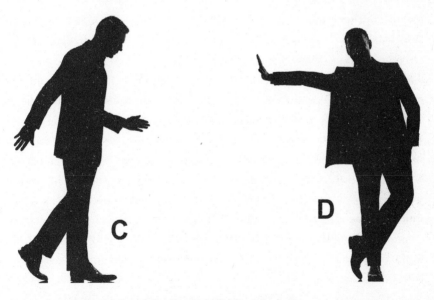

Figure 5.3: Immaterial Souls in Action.

Given this new scenario, we can ask our earlier question: How can we determine that soul X is the one that caused Charles to move his arm, and that it is soul Y that caused Denis to move his (and not the other way around)? This is an important question to be able to resolve because nobody in her right mind would want to argue that my soul could make someone else sneeze or pour himself a cup of coffee, or that his soul could make me buy a can of anchovies or grab a beer from the fridge. To each his own soul! So how could we determine that it is indeed soul X that made Charles raise his arm and that soul Y made Denis move his? Recall that our first option was to trace a continuous causal path between the bullets and their targets using a high-speed camera. In the present scenario, however, this option would be hopeless. You could film Charles and Denis all you want, using the most sophisticated high-speed camera available, but you would never be able to detect their souls. After all, souls are supposed to be immaterial and therefore invisible.

You may have noticed that in movies sometimes, like in the movie *Ghost*, souls are visible. But this is because movie directors are more concerned with dramatic effect than with scientific accuracy. If souls are visible, then they cannot be immaterial, because if they were, they wouldn't be able to absorb and reflect light. And if souls are immaterial, or nonphysical, as they are claimed to be, then they cannot reflect light and must therefore be invisible. Back to Kim's thought experiment: How about trying to find a pairing relationship between the two individuals and their souls? What could this relationship be? We could try to say that soul X is closer to Charles and soul Y is closer to Denis, but that would lead us nowhere. Again, souls have no spatial extension, and so the question of which soul is closer to which person is nonsensical. If souls are immaterial, and therefore cannot be located in space, how can one soul possibly be closer to a given body than another soul? Perhaps we can try to think of the pairing relationship in perceptual or psychological terms. Let's try perception first. We could say that soul X somehow uniquely "perceives" Charles and that soul Y uniquely "perceives" Denis (but that soul X does not "perceive" Denis and that soul Y does not "perceive" Charles). Recall that we want each soul to causally interact with only the body to which it belongs. If so, whatever *perceive* means here, it had better be a relationship that uniquely holds between one soul and one body.

So how could soul X "perceive" Charles but not Denis? What does it mean for me to perceive the house across the street but not the one that is hidden behind it? The answer is that I can perceive the house that is appropriately reflecting the photons that are hitting my retina, because I am properly positioned with respect to that house, and there is no visual obstacle intervening between me and the house. In other words, I can perceive the house that is *causing* my perceptual experience of it as a house. If so, perception too must ultimately be understood in causal terms. And this brings us back to square one. What does it mean for a soul to cause something in the physical world? We still do not know because we still do not have the immaterial equivalent of the framework that space provides in the physical world. So what does it mean for a soul to "perceive" a body? It means that the body is being "perceived" by the soul! In scientific jargon such statements are called question-begging. And when you are trying to understand something, it's preferable to answer questions and not beg them.

You can now see that using cognitive relations, such as thinking, for example, does not make things any easier. For what does it mean for a soul to uniquely "think" about Charles such that it enters into the proper pairing relationship with him but not with Denis? After all, I cannot *cause* the cell phone on my desk and the pencil next to it to move by merely *thinking* about them (if so, I would have really awesome powers). So if you want to say that *thinking* can cause something to happen in the material world, you first need a notion of *causality* in terms of which you can then define thinking. "But wait a minute!" I am sure some of you are saying. "Doesn't the act of thinking or wishing to move the pencil actually *cause* my hand to extend toward it and give it a nudge? If so, doesn't thinking have causal powers?" From the first-person perspective, it certainly feels as though conscious thoughts have causal powers, but we have learned to not trust what introspection tells us.

It may seem to us, introspectively at least, that conscious and apparently immaterial "thinking" can cause our body to do the things that it does, but this is only because our faculty of introspection is not penetrating enough to reveal to us that our conscious actions—and in fact our conscious thoughts themselves—are the result of physical activity in the brain, as I will show you in chapter 6. We will also discover that when cognitive psychologists tell us that certain mental states cause our actions, they are merely using terms like *thinking* to describe the physical activity of the brain at *a certain level of abstraction*. This means that all the actual causal work takes place between purely physical entities. Only in our descriptions do we use terms like *mental* or *thinking*, but this is merely shorthand that allows us to talk about unfathomably complex ensembles of physical processes at a much higher and more abstract level (more on this in chapter 7).

The moral of the story is that we can understand causal interactions in the physical world, like bullets hitting targets, because we have the framework of space (space-time in fact). But there is no known equivalent to the framework of space in the *nonphysical*, or *immaterial*, domain of souls, and so there is no way to even conceive of a possible mechanism for causal interaction between body and soul. A committed dualist may of course reply that this entire line of reasoning is nothing but a god-of-the-gaps argument. "Ha!" they would say. "Materialists accuse us of using the soul-of-the-gaps argument, but look at what they're doing now! Talk about the pot calling the kettle black!" But the false equivalence here borders on the grotesque. The reason there are gaps in our scientific understanding of the world is because science allows us to understand many aspects of the world in the first place. Dualism, by contrast, is one giant gap. It allows us to understand absolutely nothing, and souls are as mysterious today as they were when Plato wrote about them more than two thousand years ago. The scientific revolution took place and materialistic science got off the ground because it proved enormously useful as a method through which we could understand the world. Dualism, on the other hand, never got off the ground. There is no evidence for the detachability of body and mind, and the idea that an immaterial substance could somehow causally interact with objects in the physical world is as unintelligible today as it was when Descartes proposed the idea more than three hundred years ago. Let me now show you that the idea of an immaterial soul causally interacting with the body is not just unintelligible, it is also incompatible with what we know about physics.

THE INTERACTION PROBLEM: INCOMPATIBILITY

During the nineteenth century, physicists developed the laws of thermodynamics and things began to heat up for the soul. The first law of thermodynamics states that energy is conserved and cannot be created or destroyed. Physicist Peter Hoffman offers a very useful analogy. He asks you to imagine a situation in which someone mugs you and steals all the cash in your wallet, say one hundred dollars. So you started the evening with one hundred dollars in your wallet and now all your money's gone. But the money hasn't completely disappeared. It's simply been transferred to the thug who mugged you. Our hypothetical robber became one hundred dollars richer, but the money didn't materialize out of thin air. It already existed and was in your wallet. So, in this scenario, money was neither created nor destroyed, it was merely transferred.[6] Of course, no analogy is perfect, and thermodynamics is certainly more complicated than the simple situation I described, but the analogy is good enough for our purposes.

We can now state a standard objection to dualism based on the first law of thermodynamics. As philosopher Jerry Fodor put it:

> The chief drawback of dualism is its failure to account adequately for mental causation. . . . How can the nonphysical give rise to the physical without violating the laws of conservation of mass, of energy, and of momentum?[7]

Returning to our money analogy, an immaterial soul being able to causally interact with the physical world would be analogous to our hypothetical robber mugging you, stealing your empty wallet, and discovering that it contains one hundred dollars. Where did the money come from?! It can't simply have materialized out of thin air! This, in a nutshell, if you replace *money* with *energy*, is the problem created for the soul by the first law of thermodynamics. Where does the energy necessary to bring about a change in the brain come from?

There is, it turns out, a possible loophole for the soul. And would you believe that it involves a demon? The devilish creature was dreamed up by nineteenth-century physicist James Clerk Maxwell in a famous thought experiment. Maxwell was one of Einstein's heroes, and he is remembered today for his elegant mathematical formulation of classical electromagnetic theory. What Maxwell's demon was supposed to be able to accomplish is nothing less than violating a fundamental law of physics—the second law of thermodynamics. Now what is this second law? Imagine you are in a rush one morning and, after only two sips of your favorite brew, you leave your cup of hot coffee on the kitchen counter and head off to work. When you come back home that evening, your coffee is now at room temperature, courtesy of the second law of thermodynamics. So you go to bed, and when you wake up the next morning and make your way to the kitchen—big surprise!—you discover that your previously cold cup of coffee is piping hot again. What happened is that during the night, your morning booster was able to suck out all the heat from the ambient molecules and restore its desirable temperature—just for you.

If you've never seen this happen, it is for a good reason. This would violate the second law of thermodynamics. The second law has a directionality built into it such that interacting kitchens and hot cups of coffee will eventually reach a mutual thermodynamic equilibrium (the same temperature). However, the process cannot be reversed (unless energy is supplied). You cannot start with room-temperature coffee (and a kitchen at the same temperature) and expect that the liquid in your cup will suck up heat from the room to restore the initial temperature imbalance (your coffee being piping hot and the air in the kitchen a tad bit cooler than it was before). Again, this is an approximation and things are more complicated, but this will do for what I need to explain.

In his thought experiment, Maxwell imagined that his demon was able to "see" the molecules contained in two separate chambers, and, through a small hole in the wall separating them, the demon would be able to let in to one side all the faster moving molecules (the hotter ones) while keeping out all the slower moving ones (the cooler ones). The end result is that one of the two chambers, whose average temperature was initially the same, would become hotter than the other, in violation of the second law of thermodynamics (see figure 5.4). Here's how Maxwell himself described the situation:

> If we conceive of a being whose faculties are so sharpened that he can follow every molecule in its course, such a being, whose attributes are as essentially finite as our own, would be able to do what is impossible to us. For we have seen that molecules in a vessel full of air at uniform temperature are moving with velocities by no means uniform, though the mean velocity of any great number of them, arbitrarily selected, is almost exactly uniform. Now let us suppose that such a vessel is divided into two

portions, A and B, by a division in which there is a small hole, and that a being, who can see the individual molecules, opens and closes this hole, so as to allow only the swifter molecules to pass from A to B, and only the slower molecules to pass from B to A. He will thus, without expenditure of work, raise the temperature of B and lower that of A, in contradiction to the second law of thermodynamics.[8]

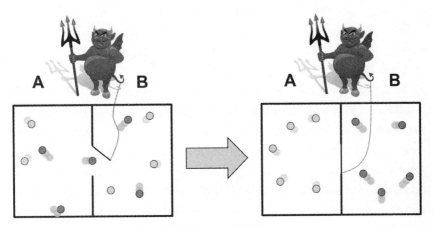

Figure 5.4: Maxwell's Demon.

Maxwell's idea can be applied to the mind-body problem and was discussed by biophysicist Harold Morowitz in a 1987 article titled "The Mind Body Problem and the Second Law of Thermodynamics," published in the journal *Biology and Philosophy*. This is how Morowitz frames the issue:

> Mind in dualist doctrine must be non-material and cannot exert force on material objects, for this would be a source of energy in contradiction to the conservation of energy. This would leave for mind the role of choosing between energetically equivalent alternatives, a perfectly plausible role within the first law of thermodynamics.[9]

First notice that Morowitz mentions the first law of thermodynamics as a problem for mind-body dualism. However, Morowitz points out, the soul could circumvent the problem posed by the first law if it could play the role of a Maxwell demon and choose between two brain states that are energetically equivalent. To make things more concrete, Morowitz considers the motor cortex, an area composed of regions of the cerebral cortex involved in the planning and execution of voluntary movements. Nerve signals take the form of electrical impulses traveling down the neurons' slender nerve fibers, the axons. The way our putative soul turned demon could operate in this context would be by selectively manipulating the ions (positively or negatively charged atoms) near the surface of a neuron's membrane, or those of the membrane itself, in order to induce the kind of electrical phenomenon that would trigger the cell to fire. This way, the soul could push our buttons and make us do the things that we do without violating the first law of thermodynamics.

You will no doubt have noticed that the argument I just described relies on one crucial postulate: the assumption that something akin to a Maxwell demon could indeed exist. The apparent paradox created by Maxwell's thought experiment haunted the imaginations of his fellow scientists until one of them, French physicist Léon Brillouin, exorcised the troubling demon in the 1950s.[10] According to Morowitz, the details of Brillouin's ghost-busting act involve black-body radiation (the kind of electromagnetic radiation emitted by an opaque, nonreflective body) as well as the fact that radiation is quantized (comes in discrete packets)—two conclusions that were unknown to Maxwell in 1871. To put things simply, Maxwell's demons have been shown to be impossible, firmly cementing the universal status of the second law of thermodynamics. Here's how Morowitz concludes his article: "Mind-body dualism thus is in direct contradiction to the second law of thermodynamics."[11]

IMMATERIAL IS AS IMMATERIAL DOES

In the previous chapter, we discovered that in spite of well-publicized claims by New Age gurus, religious apologists, and a handful of scientific naysayers, there is in fact no scientifically credible evidence for what I called the detachability of body and mind. In addition to claiming that the mind (the modern soul) can operate independently from the body, dualists also insist that the soul is *immaterial*, or *nonphysical*. But we saw that the notion of an immaterial substance, the core concept underlying the dualistic doctrine, flies in the face of what we know about modern science. Biology, through the theory of evolution, tells us that human beings are the physical outcome of a physical process operating on a set of physical raw ingredients, leaving no room or need for any additional "immaterial" substance. Physics tells us that without the conceptual framework of space-time, causal interaction between the putative soul substance and material objects is unintelligible. Physics also tells us, through the laws of thermodynamics, that souls belong to Michio Kaku's class III impossibilities.

Let us now think about what dualists could possibly mean when they tell us that the soul is immaterial. As I will show you, there is no formulation of the notion of an immaterial soul substance that isn't either trivial, and thus completely uninformative, or else utterly absurd. If this is the case, the dualistic doctrine itself is premised on conceptual quicksand upon which no solid foundation can be built. This conclusion dovetails with our earlier historical observation that the domain of the soul corresponds to the class of naturalistic phenomena that eludes our understanding. This is what we called the perimeter of ignorance. If the term *soul* is simply a name that we give to our ignorance, it is no wonder that dualism still hasn't gotten off the ground as an explanatory framework more than two thousand years after it was proposed by philosophers like Plato.

So what could a dualist mean when she tells us that her soul is "immaterial"? The first possibility would be to take the Cartesian route and claim there is a separate immaterial realm, above and beyond the material universe in which we live, and that this spiritual world isn't governed by the laws of physics. So even if scientists could one day figure out everything there is to understand about the physical universe, including a complete description of the functioning of human beings, there would remain a fundamental and unbridgeable gap in our understanding of ourselves—the domain of the mind. Let's call this *metaphysical immaterialism*. In *Life after Death: The Evidence*, New Dualist Dinesh D'Souza embraces this option and argues that "physics demonstrates the possibility of realms beyond the universe."[12] The realms that D'Souza has in mind are the other possible universes, besides our own, predicted to exist by modern cosmology. From this observation, D'Souza concludes that "there is nothing in physics to contradict the idea that we can live beyond death in other realms."

Again, D'Souza relies on flawed logic to try to make his case. To see this, consider the claim that Barack Obama is a foreign-born Muslim without a valid birth certificate, who likes to pal around with terrorists. This claim, too, is perfectly compatible with everything we know about modern physics, but that does not make it true! D'Souza seems to believe that if you can somehow vaguely point to a possible location where the soul can travel after we die (say another universe), then you have established that souls exist. This is absurd. First, you need to establish that people have souls. Then you can worry about where souls migrate after we die. As we saw, there is no evidence for the detachability of body and mind, and the idea of an immaterial soul substance that can causally interact with the body flies in the face of what we know about modern science. The fact that modern cosmology suggests there may be other universes is therefore completely irrelevant. In sum, defining *immaterial* the way Descartes did is a nonstarter because it is what led to the demise of dualism in the first place (because of evolution and the interaction problem). Besides, the claim that the human mind will never be understood in purely physical terms, because it is "nonphysical," is another instance of the argument from ignorance. How could anyone possibly know what will and will not be understood by science fifty, a hundred, or a thousand years from now?

This brings us to the second way we can make sense of notions like *immaterial* or *nonphysical*. The mind could be immaterial in the sense that it operates according to yet-to-be-discovered principles of nature. This is what New Dualists Mark Baker and Stewart Goetz seem to have in mind when they write that the soul hypothesis "might count as a minor variant of materialism rather than as true dualism."[13] In this view, notions like *immaterial* or *nonphysical* would be semantic placeholders. We call the mind *immaterial*, for lack of a better term, because it seems to be unlike anything else we know and understand. And since what we understand about nature is material or physical by definition (remember Chomsky's argument?), we call the mind *immaterial* or *nonphysical*. Let's call this *epistemic immaterialism* (because we lack knowledge in this case), to contrast it with metaphysical immaterialism. Being an epistemic immaterialist would be perfectly reasonable.

Take for example electromagnetic waves upon which so much of our modern technology relies (radios, TVs, cell phones, GPS devices, etc.). Electromagnetism was not understood in the time of Descartes (the seventeenth century), although the phenomenon was certainly part of nature then too. Since electromagnetism was unintelligible within the framework of Descartes's contact mechanics, we could then say that the phenomenon is immaterial or nonphysical because it is unlike the material phenomena that were understood at that time. Is it possible then that the human mind works according to principles that are not yet understood by current scientific theories? Absolutely. However, recall Chomsky's observation that "the material world is whatever we discover it to be, with what whatever properties it must be assumed to have for the purposes of explanatory theory" (see chapter 2). If so, notice that as soon as something that is not yet understood becomes understood and incorporated into the formulation of science, it becomes, ipso facto, *physical* or *material*. After all, what does *physical* mean other than whatever physics, or science more generally, needs to postulate to understand the physical world. Until yet-to-be-discovered principles of nature are discovered then, epistemic immaterialism simply amounts to using a semantic trick—calling something *immaterial*—to give our ignorance a more legitimate face. Needless to say, this does not bring us any closer to understanding how the mind works.

The third possibility is to define notions like *immaterial* or *nonphysical* by analogy to abstract concepts. Take for example a person's nationality. Is your nationality a material thing? Your passport certainly is, but your nationality is much more abstract and doesn't seem to be material or physical. How about numbers or information? They too would seem to be immaterial. What if we said that the soul is immaterial in the same way that nationality, numbers, or information are immaterial? There is certainly nothing illogical about making such a claim. But notions like nationality, numbers, or information are abstract concepts entertained by people, and people are physical beings. So ultimately, even abstractions seem to emerge from the physical world. As Victor J. Stenger remarks, "Not everything we talk about in our materialistic theories is itself composed of matter. An equation is immaterial; so is a distance or a number. That material objects enter into relations

and relations are immaterial does not imply that relations can exist independent of material objects. In fact, they don't. Take the matter away and you
don't have gravitational attraction or conservation of energy. Nor do you
have thought."[14]

Stenger put his finger on an important idea that we will discuss in chapter
7 when we talk about the way modern psychology views the mind. As I will
show you, the modern mind is a construct used to describe the operation of
the brain at an abstract level of analysis. So if dualists want to call the mind
immaterial because it is abstract, then they are essentially saying what cognitive psychologists are saying. And when cognitive psychologists talk about
the mind, they are really talking about the brain (at some abstract level). But
somehow, I feel that this is not what dualists want to say. What they really
want to say is that the mind is separate from the body and can operate independently from it. This is what we called the detachability of mind and body.
But calling the mind immaterial because it is an abstraction is not the same as
showing that it can operate independently from the body. In the end, this last
option also leads to an impasse for the dualist because it reduces the mental
to the physical.

We have considered three ways to define the core notion underlying the
dualistic doctrine, the idea that the mind is immaterial (or nonphysical). The
first option, metaphysical immaterialism, is a nonstarter because it relies on
a flawed argument from ignorance. Claiming that there exists an immaterial realm of mind, distinct from the physical universe that we inhabit, is
outlandish because there is no evidence for such a realm. Besides, nobody
has a crystal ball, and so nobody can seriously claim that science will never
be able to fully understand the nature and functioning of human beings
in purely naturalistic terms. At best then, metaphysical immaterialism is
wishful thinking on the part of dualists. The other two options—epistemic
immaterialism and immaterialism-as-abstraction—ultimately reduce the
mental to the physical, hardly what dualists would want to conclude. In the
end, dualism boils down to a claim about mind-body detachability for which
there is no evidence. This claim, in turn, is premised on the notion that the
mind must be immaterial, an idea that either is untenable or reduces to a
form of physicalism.

DUALISM LAID BARE

At the beginning of the book, I used a forensic analogy to explain why modern science has abandoned the soul. In my example, all the clues converged and led to the unassailable conclusion that my hypothetical suspect was guilty. Similarly, when we look at the world through the eyes of a scientist, we see that all the clues we can gather inexorably point to the conclusion that the soul must be a figment of our imagination. The domain of the soul has shrunk as scientific understanding progressed. There is no evidence for the detachability of mind and body; the notion of an immaterial substance flies in the face of what we know about modern science; and, as we will discover in the next chapter, there is overwhelming evidence supporting the materialistic hypothesis—the alternative to dualism. Under those circumstances, what can a committed dualist say? When I described my forensic analogy, I pointed out that it is always possible to look at the individual clues in isolation and fail to connect the dots. For each potentially incriminating clue, a naysayer could always concoct a story that would explain the clue away. This is the strategy of choice for the New Dualists, who seem to remain oblivious to the collective power of the evidence that so dramatically undermines their worldview.

The New Dualists never seem to be able (or willing) to connect the dots and acknowledge that the game is over. Take, for example, the argument from evolution. In their book *A Brief History of the Soul,* Stewart Goetz and Charles Taliaferro try to deflect the objection by allowing for the possibility that some nonhuman animals may have souls too:

> In reply, we do not follow Descartes with respect to nonhuman animals. We think it is reasonable to believe that many mammals have experiences, and perhaps even an awareness of themselves as distinct from others. . . . We therefore (like most dualists today) do not face the objection that we arbitrarily privilege humans in terms of the attribution of mental states. . . . We therefore do not see evolution as excluding the possession of souls by nonhuman animals.[15]

While Goetz and Taliaferro are being very kind to other animals (and are most likely right about some of them), what they offer is a false solution.

You can pick any nonhuman animal you want and find that they, just like us, are part of an unbroken physical chain. This would seem to leave dualists with only two viable options: Either everybody has a soul, or nobody does. Taking the first route (if you take the second, the game is over), you would then need to explain what capacity a soul would confer to organisms like bacteria. Language? The ability to fall in love with other bacteria? A moral compass? If we push our reasoning one step further and acknowledge the continuity between the living and the nonliving world, then we are forced to conclude that rocks have souls too. (Not that this hasn't been tried, I should add. Indeed, such forms of animism can be found in Shinto, Hinduism, and Buddhism.) Forget about bacteria, then. The questions you need to worry about now are what souls would do for rocks and how anyone could determine that rocks have souls in the first place.

Next, consider the problem posed for the soul by the laws of thermo-dynamics. In a chapter of *The Soul Hypothesis* called "The Energy of the Soul," philosopher Robin Collins takes issue with the argument against dualism based on the first law of thermodynamics (the conservation of energy). Collins argues that there are at least two ways for physical systems to interact causally without an exchange of energy. The first, Collins claims, is gravity in Einstein's theory of relativity, and the second has to do with a phenom-enon called quantum entanglement, the details of which need not concern us here. Here's how Collins puts it:

> But even this weakened version of the objection assumes both that the principle of energy conservation applies to all known purely physical interactions and that all causal interactions (or law-like connections) between events must involve an exchange of energy. The first assumption is false for the case of general relativity . . . the second assumption is false for the case of quantum mechanics. Thus, based on current physics, the energy conservation objection has little, if any, merit.[16]

I am not a physicist, but I am perfectly willing to assume that Collins did his homework and that what he is telling us about physics is correct. Even if Collins is right about gravity and quantum entanglement, these two facts are irrelevant and his general argument is misguided. What we know for sure is that Collins, or anyone else for that matter, doesn't have the foggiest idea how putative soul-body interactions work. Something else we know for sure is that whatever these putative interactions are, they cannot be *just like* quantum entanglement or gravity for the simple reason that those are *physical* phenomena and that the soul is supposed to be *nonphysical*. If Collins really thinks that the soul interacts with the body via quantum entanglement or gravity, is he ready to make psychological predictions (the domain of the soul) based on the equations of quantum mechanics or general relativity? I very much doubt it. If not, it is pointless to identify known *physical* phenomena and imply that they may represent loopholes for the *nonphysical* soul to interact with the body.

This brings us back to the original question. If we know that the soul doesn't interact with the body via the physical mechanism of gravity or quantum entanglement, then the question of conservation of energy arises again. How then does the immaterial soul interact with the body without violating the first (or the second) law of thermodynamics? A dualist might of course be tempted to reply that the soul interacts with the body in a way that is similar to quantum entanglement or gravity, maybe a nonphysical version of these mechanisms. But what would a nonphysical version of gravity or quantum entanglement even look like? Nobody has the faintest idea, and that's the general problem with dualism. Because dualists never define what they mean by *immaterial* or *nonphysical*, they never propose a concrete mechanism that could be assessed against what we know about physical laws. Remember, dualists do not tell us what the soul is; they tell us what the soul isn't. And so without a concrete proposal on the table, the New Dualists are left arguing about how many angels can dance on the head of a pin.

To be fair to Baker, Goetz, and the contributors to *The Soul Hypothesis*, Cartesian dualism isn't necessarily the view they are trying to defend. In a section of their book called "A Plurality of Soul Hypotheses," Baker and Goetz make this point explicitly:

> A problem that besets dualists in the current climate is that people talk as if there were only one kind imaginable, namely pure Cartesian dualism . . . but like any other interesting hypothesis, the Soul Hypothesis can exist in many guises, which share some basic assumptions but not others.[17]

This is all very well, but when you read *The Soul Hypothesis* it is not always clear that the authors are talking about anything other than Cartesian dualism. In his chapter on the energy of the soul, for example, Collins never specifies the kind of dualism that he has in mind, and he gives us every reason to believe that he is defending Cartesian dualism against claims that the soul would violate the law of conservation of energy. Baker himself opens his chapter by writing, "Suppose that we consider it an open question whether there is a nonphysical aspect of human beings that contributes somehow to their mental lives—something akin to the traditional notion of a soul."[18] The problem is that Baker never says anything more precise about what he takes the soul to be, and so it is not clear that what he has in mind is much different from Cartesian dualism.

To their credit, Baker and Goetz insist that the soul is a scientific hypothesis and that it should be regarded as such. And yet, in a stunning admission, they end up telling us that the conclusions they reached should be taken on faith. For one of them, we are told, the more certain truth is the existence of God. Since God is believed to be a "spirit" that is not part of the physical universe, this leaves open the possibility that there may be other "spirits," like souls. On the basis of such considerations, Baker then looks for evidence from linguistics that would support his dualistic bias. So for him "theism plus a consideration of relevant facts leads toward dualism."[19] For Goetz, the reasoning runs in the opposite direction. For him, we are told, "the more certain truth is dualism. It seems obvious to him that he cannot be simply a physical object, subject to all and only the laws of physics, given his first person experience."[20]

Of all the New Dualists that we will meet in this book, Baker and Goetz are arguably the most sophisticated. But even they can do no better than rely on faith to establish the reality of souls. For Baker, you need to believe in God to reach the conclusion that there may be other "spirits" out there—hardly a scientific argument. For Goetz, the argument is that dualism just feels obvious to him. Not exactly a scientific argument either. Baker and Goetz's book *The Soul Hypothesis* contains nine chapters on various aspects of the soul, including passionate rhetoric about what science is and how it is supposed to work. And yet, in spite of all that, Baker and Goetz end up candidly admitting that in order to believe in the soul, you either have to believe in God or be convinced that dualism is true based on how you feel. The late Christopher Hitchens put it best: "That which can be asserted without evidence can also be dismissed without evidence."[21]

THE EMPEROR'S NEW SOUL

A number of modern-day tailors, including the New Dualists we met in this chapter, have claimed that the emperor's soul is similar to his new clothes: you cannot see it, you cannot touch it, but somehow it is there. Our goal in the last two chapters, as followers of Feynman's first principle, has been to evaluate the tailors' claims using the tools discussed in chapter 2. Our conclusion so far is that there is no evidence for the existence of the soul. As we discovered earlier, the soul has shrunk over the course of history as scientific understanding progressed. The New Dualists insist that the soul must be immaterial or nonphysical, but there is no known formalism that describes the elusive soul substance. As New Dualist Mark Baker acknowledges, "There is of course no rigorous mathematical definition of the soul that we can appeal to which would be comparable to Turing's definition of computation."[22] Without a clear definition of what *immaterial* or *nonphysical* means, the New Dualists are left with a doctrine that cannot even be coherently formulated. What's more, there is no objective evidence to support the claim that the mind can operate independently from the body (what I called detachability). The idea of an immaterial soul also flies in the face of what we know about modern science. For all these reasons, the soul is also explanatorily impotent.

In the end, the soul, like the emperor's new clothes, has exactly the set of properties that it should have if it didn't exist. Moreover, as we are about to discover in the next chapter, there is a mountain of evidence supporting the materialistic hypothesis—the alternative to dualism. At the end of *The Soul Hypothesis*, Baker and Goetz wonder why the scientific community has written off the soul. Why isn't the soul hypothesis taken seriously, they ask. Baker and Goetz see two main reasons why mainstream scientists have abandoned the soul. The first, they argue, has to do with fashion—or, as they put it, "the spirit of the age" (nice play on words, I must say). The second reason they mention is that dualism has obvious religious connotations, and, according to them, most scientists are not very fond of religion. How sad and unimaginative to reduce the scientific enterprise to a game ruled by fashion and animated by antireligious sentiment. There is an obvious third possibility that Baker and Goetz do not mention. It is that mainstream scientists have abandoned the soul because reason and evidence, the tools of their trade, have compelled them to do so. I am sure that Daryl Bem and Douglas Biklen also wonder why the scientific community does not take *psi* and Facilitated Communication more seriously.

Chapter 6

LA METTRIE'S REVENGE

*Descartes and all the Cartesians, among whom the followers of
Malebranche have long been numbered, have made the same
mistake. They have taken for granted two distinct substances in man,
as if they had seen them, and positively counted them.*
—**Julien Offray de La Mettrie,** *L'Homme Machine,* 1748

n 2001, the *Journal of the History of the Neurosciences* published an article
commemorating the life of Julien Offray de La Mettrie (1709–1751),
a French physician who saw his day of reckoning following an overin-
dulgence of truffles and pheasant pâté.[1] La Mettrie's contemporaries might
have wondered whether the sin of gluttony would send a soul to hell. But
our hedonistic physician did not believe in the soul. He even expressed his
heretical views publicly in two tracts published in 1745, *Histoire Naturelle de
l'Âme* (Natural History of the Soul), and in 1748, *L'Homme Machine* (Man as a
Machine). With these arrows aimed at the priests, La Mettrie declared that
human beings are soulless automata, like animals in Descartes's system. The
ideas were not received well, to say the least. *Histoire Naturelle de l'Âme* was
condemned to be burnt, and La Mettrie found it wiser to leave France to
find refuge in the Netherlands.[2] The reason he is remembered in the pages of
neuroscientific journals today is that history proved him right.

One of history's notable ironies is that Descartes and La Mettrie reached
their antithetical conclusions while both undergoing the effects of intense
heat. In early November 1619, after spending a day of meditation in a stove-
heated room following a night of wild dreams, Descartes had the epiphany
that led to the foundations of his new philosophy.[3] La Mettrie's radically
materialistic vision occurred to him while he was suffering from a burning
fever in a hospital bed in Freiburg. Despite his condition, La Mettrie was
lucid enough to notice the chaotic flow of his thoughts. From that, he

concluded that the mind is nothing but an extension of the body and that Descartes's soul is a fiction. This insight formed the central conclusions of *Histoire Naturelle de l'Âme* and *L'Homme Machine*.[4]

In the last two chapters, we reviewed two lines of reasoning that have led to the present scientific consensus away from the soul. The first is that dualism has failed to make good on the self-imposed burden of proof that befits any scientific hypothesis. As we discovered in chapter 4, there is no credible evidence supporting the detachability of mind or the existence of the mysterious soul substance. Modern science also tells us that dualism flies in the face of established conclusions in the biological and physical sciences. The final nail in the soul's coffin comes from the mountain of evidence supporting the materialistic hypothesis we defined in chapter 2.

If materialism is indeed true, we can ask ourselves what we would expect the world to look like. If the mind is simply a description of the brain at a more abstract level, we should expect a number of things to follow. First, because the brain is material and thus divisible into parts, the mind should be divisible as well, at least in principle. On the other hand, if the mind is immaterial, and therefore indivisible, as Descartes believed, then we should not be able to break it up into pieces. Let's call this *the divisibility of the mind*.

Second, if consciousness can remain fully operational after the brain has turned into mulch, as believers in the afterlife contend, then only *partial* damage to the brain should have no effect on consciousness. By contrast, if consciousness is caused by the operation of the brain, then damaging the brain should also impact consciousness. Let's call this *the fragility of the mind*.

Third, if what gives rise to the mind and voluntary behavior is triggered by an influx of soul substance, then it should not be possible to generate conscious experience or behavior by simply poking at the brain. If, on the other hand, the activity of the brain is what causes the mind, then we should be able to push the mind around by directly stimulating the brain. Let's call this *mind control*.

Fourth, if our mental world is the result of an invisible soul substance, then it should not be possible to "read" people's minds by simply looking at their brain. Let's call this *mind reading*.

Finally, if our abilities to think, reason, make decisions, and display insight and creativity cannot come from the properties of physical matter alone, then it should not be possible to build physical devices that display these mindlike properties. Let's call those *intelligent machines*.

The divisibility and fragility of the mind, mind control, mind reading, and intelligent machines are all a reality today. (Some of these conclusions have even been known for a quite a while, as we saw in chapter 1.) These phenomena make perfect sense in the context of a materialistic view of mind. Their combined weight thus lends powerful support to the conclusion that the mind and the brain are two sides of the same coin. For a committed dualist, however, these properties are much more problematic, especially when considered collectively. Why should the soul be divisible into parts? Isn't it supposed to be immaterial and therefore indivisible? How exactly does one cut a ghost into pieces? If damage to only parts of the brain can make you lose your ability to see, think, or feel, then how can all these abilities remain intact when your whole brain is completely kaput? And if consciousness can be pushed around by physically stimulating the brain, then the soul substance is not necessary to generate the mind. Why then do we need the soul in the first place? Wouldn't Descartes be forced to conclude that my iMac, my iPhone, and the GPS device in my car have souls?

This is the kind of converging evidence that so impresses scientists, and rightly so. *Consilience* is the term used to describe the cumulative force of unrelated sources of evidence, often coming from different disciplines, to support strong conclusions. Most established scientific knowledge takes this particular form. Coming back to my *CSI: Miami* analogy: Everybody knows about the power of consilience. If the accused does not have an alibi, if the motive for the crime is transparent, if the suspect has been positively identified by eye-witnesses and video surveillance equipment, if there are traces of the victim's blood on his clothes, if he can be placed at the crime scene at the right time, and if DNA evidence from other bodily fluids links him to the victim, what are the odds that he would actually be innocent?

How then do the New Dualists respond to consilience? How do they explain the combined lack of evidence for the existence of the soul, the sheer implausibility of dualism in light of what we know about modern science, and the mountain of evidence supporting materialism? Well, they simply don't—they just can't. Instead, they try to instill doubt, they make unfalsifiable claims about soul-brain interaction, they complain that materialism hasn't explained everything yet or that it is deeply flawed, and they maintain that future developments will vindicate their views because common sense is on their side. They insist that a complete physical account of the mind is a fool's errand, that it is simply impossible. In short, they remain in denial.

Since the formulation of Descartes's substance dualism in the seventeenth century, "soul science" has made no progress whatsoever. It is what philosophers call a *degenerative* research program. There isn't a single soul equation or other piece of formalism apart from trivial analogies to radio sets or vibrating guitar chords[5]—all physical devices, mind you—not one scintilla of objective evidence, not one positive result that dualists can claim. Souls are supposedly immaterial, they are objectively undetectable, devoid of any explanatory force, but somehow, dualists insist, they are there. Soul advocates themselves still do not have the foggiest idea what the soul is, how it interacts with the body, and how it gives rise to the mind. And so naturally, since dualism has nothing to show for itself, soul advocates criticize materialism instead. Sometimes, the best defense is indeed a good offense. By contrast, materialistic mind science has progressed by leaps and bounds since the seventeenth century—an important fact that dualists are very happy to sweep under the rug. Materialism is what philosophers of science call a *progressive* research program.

Biologists, neuroscientists, linguists, computer scientists, and psychologists have made discoveries in the last few decades alone that would have blown Descartes's socks off (or whatever similar thing people wore in the seventeenth century). Materialism, just like dualism, is not exempt from the strictures imposed by the burden of proof. In the pages to follow, I would like to show you that there are very good reasons indeed to believe that materialism is true and that Julien de La Mettrie was a visionary. If there is no positive evidence supporting dualism, if modern science renders the doctrine untenable, if it is explanatorily impotent, and if all the evidence points toward materialism instead, then it is time to acknowledge what reason is trying to tell us—there is most likely no soul.

DECONSTRUCTING THE SOUL

When I woke up this morning, I felt exactly like the same person who went to bed the night before. My consciousness was temporarily interrupted by sleep (I do not recall having any dreams), but when I woke up, nothing had changed. I am the same *me*, the same *I*, the person who had salmon for dinner the night before, the French student who immigrated to the United States, the professor of psychology at Rutgers University, the author of *The Soul Fallacy*. This *I* is also the private theater where the data from my senses come to life in a perfectly orchestrated show full of lights, sounds, and tastes. The familiar aroma of coffee, the sweet taste of blueberries, the festive sight of that Christmas tree we haven't yet recycled, the gentle warmth of the water spraying down on me from the showerhead. The unity of conscious experience is a ubiquitous fact of life central to the human condition. Thinkers like Plato and Aristotle already mused over it.[6] The unity of the self is also what led Descartes to propose that the mind and the body are not cut from the same cloth, that the *I*, the *self*, the *soul* is indivisible and that it must emanate from a radically different substance. This is how the neurophysiologist and Nobel laureate Charles Sherrington (1857–1952) describes the unity of the self:

> This self is a unity. The continuity of its presence in time, sometimes hardly broken by sleep, its inalienable "interiority" in (sensual) space, its consistency of view-point, the privacy of its experience, combine to give it status as a unique existence. . . . It regards itself as one, others treat it as one. It is addressed as one, by a name to which it answers. The Law and the State schedule it as one. It and they identify it with a body which is considered by it and them to belong to it integrally. In short, unchallenged and unargued conviction assumes it to be one. The logic of grammar endorses this by a pronoun in the singular. All its diversity is merged into oneness.[7]

But Descartes was wrong. It certainly feels as though the mind is indivisible, but it isn't. One of the most spectacular demonstrations of the divisibility of the mind comes from a group of people we met in chapter 2. The peculiar behavior of split-brain patients teaches us that physically bisecting the brain also divides the mind. As we saw earlier, split-brain patients cannot name an object presented in their left visual field because the information can no longer travel from the "mute" right hemisphere to the "speaking" left hemisphere. The result is that the left hemisphere isn't aware of the information that was presented to the patient, but the right hemisphere clearly is, since it can guide the patient's left hand to select the relevant object. It is as if two minds, unaware of each other, inhabit the same person.

The presence of two "minds" can also be detected when split-brain patients perform certain actions. Imagine being shown two figures, one in each visual field, and being asked to draw them with both hands simultaneously. Individuals with intact brains can do so as long as the figures are identical or mirror images of each other. However, when the figures yield incompatible spatial maps (see lower half of figure 6.1), people with intact brains can no longer perform the task, because their unified self cannot handle conflicting motor programs. Split-brain patients, however, experience no difficulty with that task and outperform normal controls. In split-brain patients, the two different "selves" can control both hands without having to worry about each other.[8]

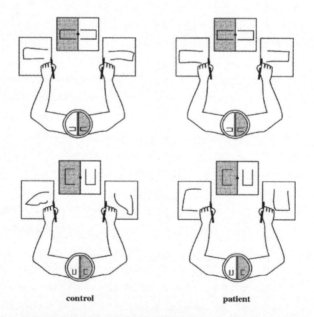

control   patient

Figure 6.1: Split-brain patients in action. Image from Michael S. Gazzaniga, "Cerebral Specialization and Interhemispheric Communication," *Brain* 123 (2000): 1299. Used with permission from Oxford University Press.

The fascinating behavior of split-brain patients demonstrates the divisibility of the mind. But can we conclude from these observations that consciousness itself is divided in these patients? Recall the analogy from chapter 2 in which we equated the conscious mind with the tip of an iceberg and the unconscious mind with the iceberg's body. The behavior of split-brain patients demonstrates that their unconscious mind has been divided. In the drawing task I just described, the patients act as though independent minds or selves control each hand. But do split-brain patients also report feeling divided? Remarkably, this does not seem to happen. These people generally do not have any clue that something so radical happened to their brain.[9] Does that mean that conscious awareness—what we called the unity of the self—is invulnerable? Hardly.

In his book *Life after Death: The Evidence*, Dinesh D'Souza asserts that he has found two features of human nature that appear to be independent from matter: consciousness and free will. He adds that because these two capacities are not constrained by physical law, they cannot perish or be destroyed.[10] I wonder what D'Souza tells his dentist, then. That he doesn't need anesthesia because consciousness is not governed by physical law and therefore cannot be extinguished, even temporarily? If science has not (yet?) reached a full understanding of consciousness and free will, then how can D'Souza confidently assert that these capacities are not governed by physical law, especially when the facts on the ground point in exactly the opposite direction? To illustrate this conclusion, let me tell you the tale of three fascinating syndromes, straight from the X-files of cognitive neuroscience, demonstrating that the capacities New Dualists are so fond of touting as other-worldly can be affected by physical changes in the brain, just as materialism would predict. This is what I have called the fragility of the mind.

In 1908, German neurologist Kurt Goldstein reported the case of a fifty-seven-year-old woman who claimed that her left hand had a mind of its own and that it performed actions she did not want it to do. On one occasion, the uncontrollable hand even grabbed the woman's throat and tried to choke her. The poor lady believed her hand was possessed by an evil spirit. The experience of conscious will, something we will explore in the next chapter, is part of everyday life. If I want the glass of water that is sitting next to me on my desk, all I have to do is consciously tell my hand to grab it, and this will be followed by the appropriate action. How disturbing must it be, then, to have your own hand perform a meaningful action without it being accompanied by the usual feeling of authorship! And yet this is precisely what happened to Kurt Goldstein's patient. After her death, an autopsy revealed that she had suffered multiple strokes and that her corpus callosum was also damaged.

The strange condition resulting in the loss of voluntary control described by Goldstein is now called the Alien Hand Syndrome (AHS). This movement disorder was immortalized by Stanley Kubrick in his movie *Dr. Strangelove*, in which the main character's erratic right hand would at times raise into a Nazi salute. AHS typically follows from lesions in the medial frontal lobes of the brain and the corpus callosum, but it can also result from strokes occurring farther back in the brain, as well as degeneration of brain structures.[11] In his fascinating book *Altered Egos: How the Brain Creates the Self*, psychiatrist and neurologist Todd Feinberg tells us about one of his patients, Stevie, a man in his late sixties who suffered from a stroke, and, as a result, developed an alien right hand. Unlike the split-brain patients I described earlier, people with AHS are fully aware that something is wrong. They report that they can no longer voluntarily control their alien hand. This is how Stevie described his experience to Dr. Feinberg:

> This is the bad boy [indicating the right hand], and this is the good boy [pointing to the left]. I control the good boy. I can't control the bad boy. That's been the problem. This hand over here—which is the good boy— is very dominant over the bad boy—this one here. Because he is my boy [pointing to himself]. You understand that? He does my wishes . . . and "bad boy" goes off on his own.[12]

The sense of conscious will that D'Souza assures us is utterly magical is in fact anchored in the physical brain. Damage a patch of brain tissue, and you'll pierce a hole right though the soul. Consciousness is fragile. The unified sense of self that I described earlier is also vulnerable, as amazing as that sounds, and can be wiped out in people suffering from Cotard's Syndrome (CS). This rare condition was named after French neurologist Jules Cotard, who described it for the first time in 1880. In a recent overview of the syndrome, the authors report on two cases from their own hospital.[13] The first involved an eighty-eight-year-old man who was admitted for treatment after a severe bout of depression. The poor fellow was convinced he was dead, and he even fretted over the fact that he was not yet buried! The second case was that of a forty-six-year-old woman who suffered from bipolar disorder. She was convinced that she had no "self," that she was only a body without meaningful content. As morbid as this may sound, she was also convinced that she no longer had a brain, that her intestines had vanished, and that her entire body was translucent. She refused to bathe or shower because she thought the water would dissolve her body and she would disappear down the drain.

CS is a perplexing condition that was described more than a century ago and presents with the frightening nihilistic delusions I described. Imagine being convinced that you are just an empty shell, a soulless body so fragile it may dissolve in water. Researchers who have studied the brains of people suffering from CS believe that lesions or abnormalities in the frontotemporoparietal circuitry may be responsible for the condition. However, CS is still poorly understood and it remains a matter of debate whether the unusual symptoms warrant postulating an independent syndrome or whether they follow from other, more general, underlying conditions.[14] Nevertheless, the existence of Cotard's Syndrome serves as a powerful reminder that our unified sense of self is a fragile construction of the brain.

The third condition I have chosen to describe to illustrate the fragility of the mind touches a nerve, so to speak, because it involves the quintessentially subjective experience of pain. Like the Alien Hand Syndrome and Cotard's Syndrome, this remarkable condition appears to come straight from the script of a science fiction movie. One of the scenes in the second *Terminator* movie features Sarah Connor and her son trying to remove bullets lodged in the Terminator's back. (Arnold Schwarzenegger's character is a good guy in the second movie.) At one point, the son asks the robot if it hurts when he gets shot. The Terminator responds that he senses injuries and that the data could be called pain. The cyborg notices when a bullet penetrates his body, but he does not suffer from it.

In *Altered Egos*, Feinberg describes meeting a patient who bore a striking similarity to the robot played by Schwarzenegger, a condition known as pain asymbolia. When John, a middle-aged electrician, walked into Feinberg's office and shook the doctor's hand, Feinberg noticed chemical burns on his right hand so severe that they had seared the skin away and exposed the muscles and bone. When Feinberg asked John about his hand, the man replied that he had done some plumbing work in his house and received "little burns." Despite the severe wounds, John had applied no bandage, and he casually used his injured hand to shake the doctor's. John's wife, Joyce, explained to Feinberg that her husband sustained the burns from chemicals he came in contact with as he tried to fix the kitchen drain. Joyce said she noticed blood and pus on John's clothes when she sorted out the laundry. John's injuries had festered for days, but he went about his business unconcerned.

John otherwise appeared to be perfectly normal, an impression that was confirmed by the result of a battery of cognitive tests. However, Feinberg learned that John had fallen from a scaffold and hit his head a couple of months earlier, an accident that sent him to the hospital for several weeks and put him out of work for a while. When Feinberg ordered an MRI scan of John's brain, he discovered that large portions of his frontal and parietal lobes had been destroyed by the accident. The subjective sensation of pain, it turns out, is also generated by the brain. This shows that despite the inherently subjective nature of conscious experience—which philosophers call *qualia*—such experiences are also under the control of our physical brain.

John's condition may be rare, but a simple trip to the dentist reminds us that we all know what it is like to experience pain asymbolia. In her book *Touching a Nerve: The Self as Brain*, philosopher Patricia Churchland asks us to think about what happens when your dentist freezes a nerve in a tooth and your "soul" ceases to feel the pain. Neuroscience offers a detailed explanation based on a precise set of well-understood mechanisms, fully compatible with everything else we know about science. (Remember consilience?) The local anesthetic the dentist injects into your jaw acts on the sensory neurons near the tooth and prevents them from sending pain signals to your brain. The details of the mechanism are also known: When a neuron fires, sodium ions get pumped in and out of the cell, thereby generating a nerve impulse—the signal sent to your brain. Novocaine temporarily blocks sodium channels and, therefore, prevents the cells from firing.

Now how does the story work for a dualist? How can a physical substance like novocaine act on the immaterial soul? What is the mechanism here? The answer is that nobody has the faintest idea. There is no soul-based explanation that even comes close to matching the details of the materialistic account. As we saw earlier, our ancestors invoked souls to describe phenomena they did not understand. Some of the greatest scientists also invoked God when they reached the limits of their own understanding. But once nature delivers its secrets to science, God and the soul become superfluous. A dualist might, of course, reply that neuroscience still doesn't understand why or how subjective experiences—qualia—arise in the brain. But so what? Do dualists understand that any better? I don't think so. Try this as an explanation: The soul gives rise to qualia. Do you now understand subjective experience any better?

The story of split-brain patients, the Alien Hand Syndrome, Cotard's Syndrome, and pain asymbolia represent some of the most impressive demonstrations that our unified sense of self, our sense of conscious will, and our subjective feelings of pain—phenomena so dear to dualists—can all be affected by brain damage. But the list doesn't stop there. Your memory, your ability to talk, and your personality can be wiped out by brain damage. People who suffer from asomatognosia will assure you that part of their body, say their left arm, does not belong to them. In anosognosia, patients are convinced that a paralyzed limb is perfectly functional. The Capgras delusion is a condition in which patients sincerely believe that their loved ones have been replaced by impostors. Individuals who suffer from Fregoli syndrome hold the delusional belief that they are persecuted by a person who can take the appearance of different people. All these conditions result from damage to different areas of the brain. The allegedly indestructible soul is very fragile indeed. In light of such evidence, how can anyone believe that the mind will continue to function when the entire brain has given up?

MIND CONTROL

For more than fifteen hundred years—from Galen, the great Roman physician, to Descartes in the seventeenth century and beyond—theories of the nervous system involved a mysterious substance called the animal spirits. For the ancients, the animal spirits were weightless and invisible. For Descartes, they were liquids flowing through hollow nerve tubes, set in motion by the action of the soul in the pineal gland. It wasn't until Luigi Galvani began experimenting with frogs and electricity in the late eighteenth century, and his nephew Giovanni Aldini publicly electrified decapitated criminals to make them sit upright, that the theory of animal spirits began to lose steam.[15] In the nineteenth century, scientists discovered that nerves are made out of cells, and they began to decipher the electrochemical language of the nervous system, a quest that continues to the present day with understanding pushed down to the molecular level. The animal spirits have been fully exorcised.

If the nervous system gives rise to our mental lives, as predicted by materialism, then it should be possible to push the mind around by physically stimulating the brain. In the first half of the twentieth century, the pioneering work of Canadian neurosurgeon Wilder Penfield showed just that. Penfield treated patients who suffered from severe epilepsy by destroying the areas of brain tissue responsible for the seizures. Before operating, however, Penfield stimulated the brains of his patients with electrical probes while they were still conscious, in order to see how they would respond. This ingenious procedure (still in use today) allowed Penfield to evoke a wide range of behaviors from his patients and to create detailed functional maps of the motor and sensory cortices. More importantly, this allowed him not to blindly destroy brain tissue and therefore to minimize potentially devastating side effects.

And yet, as modern-day dualists like to remind us, Penfield was part of the Cartesian club too, and he believed that the mind is immaterial. As Stewart Goetz points out, Penfield was able to cause his patients to move their limbs or vocalize by stimulating their brains, but these actions were not accompanied by the usual feeling of authorship. Penfield's patients reported experiencing no intention or desire to act. They told Penfield that *he* made them do it. As Penfield explained in his book *The Mystery of the Mind*, "There is no place in the cerebral cortex where electrical stimulation will cause a patient . . . to decide."[16] Goetz concludes that this is because choices are undetermined events that belong to a different realm altogether. If intentions originate in the immaterial soul, then they surely cannot be evoked by something physical like an electrical current. Or so Goetz would have us believe.[17]

In 2009, a group of French neuroscientists published a study in the journal *Science* that directly refutes Goetz's interpretation. The team of researchers showed that it is possible to evoke "intentions" by electrically stimulating the posterior parietal cortex of their patients. The patients themselves reported a "desire" or a "will" to move and said things like "I felt a desire to lick my lips" or "[I experienced a] will to move." Crucially, however, these intentions were not followed by actual actions, nor were they accompanied by activity in the relevant muscles. As the authors themselves explain, they were able to evoke "pure intentions." When the surgeons increased the intensity of the electrical stimulation to the same regions, patients not only reported an intention to move, but they were also convinced that they had performed a movement that in reality never took place! By stimulating the premotor cortex, the team was also able to trigger complex movements that the patients themselves were not able to consciously detect. In their article, the authors explicitly introduce their findings by observing that they fly in the face of dualism.[18]

Another technique used by cognitive neuroscientists to stimulate the brain and push the mind around is transcranial magnetic stimulation (TMS). TMS is a noninvasive technique in which brief or repetitive (rTMS) magnetic pulses are applied to a person's head. The magnetic field passes through the scalp and induces electrical currents in the brain that can disrupt normal activity. Using this technique, neuroscientists can work with healthy subjects to create "virtual lesions" in "virtual patients." Needless to say, the effects are only short-lived, and participants do not end up with permanent brain damage! Early attempts to stimulate the brain using magnetic fields were taking place as far back as the beginning of the twentieth century, but the first successful use of TMS was reported in 1985 by a team of scientists working in Sheffield.[19] Since then, TMS has been used as a tool to address basic scientific questions in cognitive neuroscience, including the study of motor learning, working memory, language production, and visual perception and attention. TMS has also been used in clinical settings as an alternative antidepressant therapy.[20]

In recent years, TMS has produced results that would have called for more burnings in the times of La Mettrie because they touch on a sphere of human experience that religion has been claiming as its turf for centuries—the domain of morality. In chapter 4, we saw how New Dualist Dinesh D'Souza asserted that the existence of moral values defies the law of evolution and can only be explained if we assume that human beings inhabit two worlds: the natural world described by science and a divine realm of cosmic justice from which our sense of morality directly flows.[21] The students we tested in the soul-belief experiment I described in chapter 2 also believed that one of the main functions of their soul was to give them a moral compass. The materialistic alternative to these soul claims is that our moral sense is the result of biological evolution, and that, like every other human capacity, it has a physical basis in the brain. This leads to the very interesting prediction that by tampering with the brain one should be able to push people's moral judgments around.

Imagine a scenario in which two women—let's call them Grace and Sarah—are taking a tour of a chemical plant. After a while, they decide to take a break and gather around the coffee machine. Sarah pours herself a cup of coffee and asks her friend Grace for some sugar. Grace notices a white powder by the coffee machine that looks like sugar. The powder is indeed sugar. But it is in a container labeled "toxic," and therefore Grace believes that it is toxic. She nevertheless puts some of the powder in Sarah's coffee. Sarah drinks the coffee and is fine because the powder is only sugar. If I were now to ask you to judge Grace's behavior, would you say she committed an act that was morally reprehensible? I hope you would! Our moral intuitions tell us that Grace did something evil because she believed the white powder was a toxic substance (even if it turned out to be harmless). This intuition is reflected in a basic principle of criminal law that "the act is not culpable unless the mind is also guilty."

As this simple example illustrates, judging the moral character of an action requires the ability to reason about people's mental states. What was Grace thinking when she put the white powder in Sarah's coffee? What did she really believe? Research in cognitive neuroscience has led to the conclusion that the neural basis of mental-state attribution involves a network of brain regions including the right temporoparietal junction (RTPJ), an area of the brain where the temporal and parietal lobes join. This conclusion leads to an intriguing possibility. What would happen if healthy participants were asked to produce moral judgments in response to scenarios like the one you just read, while TMS was applied to the RTPJ? Would it really be possible to affect something as intangible as moral judgments by applying electrical currents to the brain?

This is precisely the question that a group of researchers at Harvard and MIT set out to investigate in a recent study. In a first experiment, healthy participants received TMS to the RTPJ for twenty-five minutes before being asked to judge scenarios like the one you just read. TMS was also applied to a control area of the brain not known to be involved in mental-state attribution. Amazingly, RTPJ stimulation led participants to judge actions like Grace putting a substance she believed to be toxic in her friend's coffee as *more morally permissible* compared to judgments by the same individuals in the control condition or by participants who received no TMS at all. These results were replicated in a second experiment, in which participants received more powerful—but very brief—TMS bursts right after they finished reading the scenarios and were about to be asked to produce moral judgments. The results were the same. The moral of the experiment isn't that we can turn otherwise-normal individuals into temporary psychopaths, but that the basis of human morality is to be found in the activity of the living brain.[22]

RISE OF THE MACHINES

In *The Terminator*, machines become self-aware, rebel against their creators, and send the robot played by Schwarzenegger back through time from the year 2029 to alter the course of history. Today, humanlike robots are still part of science fiction. But who knows what the future will bring. According to futurist Ray Kurzweil, the point at which artificial intelligence will surpass human capabilities is only a few decades away. Whether realistic or not, such pronouncements follow from a more general prediction of materialism. If intelligence is a property of organized matter and does not depend on a mysterious soul substance, then building machines with humanlike mental capacities should be possible, at least in principle. What's more, if thoughts and consciousness arise from the activity of brains, materialism also predicts that it should be possible to read people's minds by decoding patterns of brain activity. Intelligent machines and mind reading are a reality today, and they have given mainstream scientists yet another reason to abandon dualism.

A reasonably good definition of *intelligence*, which also has the merit of being easy to convey, is that it is the ability to achieve goals in the face of obstacles.[23] If the Terminator's artificial brain led it to stand in the middle of a crowded intersection and randomly open fire on the off chance that Sarah Connor will happen to be around and cross paths with one of the bullets, we would all agree that the futuristic robot is pretty dumb. What makes Schwarzenegger's character so chilling is that there is method to its madness. In order to achieve its lethal goal, the cyborg impersonates human beings, extracts information from databases, steals a police car, and systematically kills all the Sarah Connors in the Los Angeles area. The robot's no dummy. Fortunately, intelligence usually manifests itself in less grisly ways, and battles between human minds can also take the form of harmless pastimes.

One of the oldest and most popular board games is the game of chess. In his song "One Night in Bangkok" (my wife often tells me that I am stuck in the 1980s), the English singer Murray Head describes the game as the pinnacle of intelligent activity. In the late eighteenth century, a few decades after the publication of La Mettrie's *L'Homme Machine*, Wolfgang von Kempelen (1734–1804) built a chess automaton called the Turk in an effort to impress the empress of Austria. The Turk was in fact a hoax: a human player hidden inside the machine could operate it, giving the illusion that the automaton could play chess all by itself. In the late 1990s, IBM built its own version of the Turk, a computer called Deep Blue. Unlike its predecessor, the IBM machine was no hoax. Much to the dismay of its detractors, Deep Blue was able to defeat then world champion Gary Kasparov in a 1997 match. (The year before, Kasparov had defeated an earlier version of Deep Blue, which had then been extensively reprogrammed for the rematch.) For the first time in human history, a mere "machine" was able to surpass the best human mind at an activity that epitomizes intelligent behavior.

A little over a decade after Deep Blue's success, IBM built Watson, another artificially intelligent machine, to try its hand at the popular quiz show *Jeopardy!* In the game, the clues are given in the form of answers, and contestants must provide their responses in the form of questions. In 2011, Watson competed against former *Jeopardy!* champions Brad Rutter and Ken Jennings, both real human beings. The computer won the first prize, and the epic contest was immortalized on YouTube. After watching Watson perform for a few minutes, I instinctively granted the machine possession of a first-rate mind. And yet, as unreal as it may sound when you watch it perform, Watson is nothing more than an assembly of electronic components designed to implement the latest techniques in natural language processing, machine learning, automated reasoning, information retrieval, and knowledge representation. Since its public debut on *Jeopardy!*, Watson has also been used in the field of healthcare, where it is helping nurses and doctors generate diagnostics and treatment options.[24]

In touting the prowess of artificially intelligent systems like Deep Blue and Watson—or your smartphone for that matter—I certainly do not mean to imply that these devices are conscious or that they operate in exactly the same way as the human brain does. Artificial intelligence also remains limited compared to the natural kind (for how long though, one feels compelled to ask). Nevertheless, the lesson could not be clearer: What we chauvinistically call intelligence, insight, and creativity, all properties once attributed to the soul, can now be programmed into computers. I cannot help but wonder what Descartes and La Mettrie would have thought if presented with such technologies.

Dualists would no doubt argue that the soul's real function is to give us consciousness, our private first-person perspective, and that until we are able to build sentient machines, artificial intelligence does not pose a real threat to the doctrine of the soul. Maybe so, but the reality of artificial intelligence is certainly pushing dualists into a corner. Remember that for Plato and Aristotle, the soul was a life soul. In the seventeenth century, Descartes showed that Aristotle's nutritive and sensitive souls had outlived their necessity. Only the rational soul remained. But one by one, properties of the rational soul are being shown to be programmable into machines. So what exactly is left for the rational soul to do? Consciousness, we are told. Fine. But if consciousness takes place in the immaterial medium of the soul, then it should be utterly inaccessible to materialistic neuroscience, a prediction that has already been falsified. As we just saw, neuroscientists can evoke pure conscious intentions by electrically stimulating the brain.

And there is more: Machines can be programmed to behave intelligently, and they can also be used to look directly into "the soul." In the 1990s, neuroscientists began using a new, noninvasive brain-imaging technique called functional magnetic resonance imaging, fMRI. This tool allows researchers to monitor the activity of the brain by measuring changes in blood flow that are closely linked to neural activity. The results can be represented graphically by color-coding the strength of activation in different areas of the brain. Using fMRI, investigators can correlate mental activity with brain activity and ask which areas of the brain will "light up" when subjects read sentences, perform a memory task, or engage in moral reasoning. These localization studies allow neuroscientists to create detailed maps of the thinking brain. Consider now what would happen if it were possible to reverse the process. The idea would be to try to recover the content of a person's consciousness by looking at patterns of brain activity.

In one study, for example, a group of researchers had subjects view pictures of human faces, cats, and an assortment of man-made objects (houses, chairs, scissors, shoes, and bottles), as well as scrambled images (as a control), while their brains were being scanned in an fMRI machine.[25] The million-dollar question was whether each category of stimulus would evoke a distinct pattern of brain response. If so, it would be possible to infer from the pattern of brain activity the type of picture that the subjects were seeing in their mind's "I." Let me look at your brain and I'll tell you whether you were viewing a face or a house, a shoe or a pair of scissors, a bottle or a chair. The results revealed that overall, the categories being viewed by the subjects were correctly identified 96 percent of the time on the basis of patterns of brain activity alone. If conscious perceptions take place in the immaterial soul, such feats of mind reading should simply be impossible.

Brain-imaging techniques can be used to obtain snapshots of the content of consciousness (house vs. face, shoe vs. bottle). But what about dynamic patterns? After all, we experience consciousness as a stream of thoughts, ideas, and impressions all free-flowing through our mind's "I." Could fluctuations in conscious awareness also be detected using brain-imaging techniques? To answer this question, cognitive neuroscientists look to a nifty perceptual phenomenon called binocular rivalry. Normally, each eye sees a slightly different image of the world and our brains combine these two images into a single, coherent representation. But when each eye is presented with a sufficiently different image at the same perceived location, something very interesting happens. Common sense would tell us that the brain will fuse the two images into a single conscious representation, but instead one image dominates conscious perception for a few seconds, until the other one takes over for another few seconds. This alternating process continues back and forth as long as one cares to keep looking. What's really neat about binocular rivalry is that the stimulus does not change; only our conscious perception of it changes.

Using binocular rivalry, researchers can ask their subjects to report on their fluctuating conscious experience by, say, pressing a button corresponding to each image when a perceived switch takes place. This way, neuroscientists can obtain a timeline of the person's fluctuating conscious awareness. At the beginning of the experiment, the subjects report seeing red vertical lines; three seconds later, blue horizontal lines are perceived, and so on. By now, I am sure you can anticipate the next question. Can patterns of brain activity detected by an fMRI scanner be used to predict the temporal sequence recorded for each subject? In other words, can someone's stream of consciousness be read by a machine? Recent work on "brain reading" shows that this is indeed possible.[26]

Better still, researchers can even read the contents of the unconscious mind using the "brain reading" techniques I just described. I already told you about the phenomenon of masked priming in chapter 4. A stimulus and a mask are presented in rapid succession, such that the subject in the experiment reports not being consciously aware of what the stimulus was. However, as we saw, other measures clearly indicate that the stimulus was unconsciously registered and that it can affect the subject's subsequent behavior. Using a slight variation of this paradigm, researchers can present masked stimuli to their subjects while their brains are being scanned, confirm that the subjects have no conscious awareness of the stimuli, and then look at the patterns of brain activity to see if they can be used to infer the nature of the masked stimuli. This can be done, for example, by briefly presenting a pattern of lines with a particular orientation followed by a mask that will render the orientation inaccessible to consciousness. Patterns of activity in the primary visual cortex of the subjects can then be used to decode the orientation of the "invisible" patterns of lines.[27]

As impressive as the feats of current mind-reading technology might be, we must also not forget that important limitations still stand in the way of what can be accomplished. As the researchers themselves point out, a general "brain reading device" still belongs to the domain of science fiction.[28] For the time being, only relatively simple thoughts can be deciphered using brain-imaging techniques, and these results must be obtained under tightly controlled experimental conditions. The prohibitive cost and limited portability of fMRI scanners impose another set of limitations on practical applications. What's more, the decoding algorithm used in "brain reading" almost invariably needs to be trained for each participant individually and for a fixed set of stimuli. This raises important questions of generalizability that will need to be resolved as well. Nevertheless, this research provides spectacular proof for the feasibility of mind reading. Current limitations notwithstanding, these developments have already led to the commercialization of neuromarketing and lie-detection products.[29] Meanwhile, in the pages of research articles, neuroscientists are beginning to warn us about the ethical considerations associated with their newest brainchild. We are in for interesting times. For us, the important lesson is that if the contents and temporal dynamics of the mind can be deciphered by analyzing patterns of physical activity in the brain, it is difficult to maintain that conscious awareness takes place in the immaterial medium of the soul.

LEAVE IT TO DUALISM

What do the New Dualists respond when presented with the evidence we discussed in this chapter? Strategies vary, but a small number of tactics are usually deployed. One approach consists in sweeping the evidence under the rug and rejecting materialism altogether. This is the path taken by Mario Beauregard and Denyse O'Leary in their book *The Spiritual Brain: A Neuroscientist's Case for the Existence of the Soul*. These New Dualists not only question materialism, but they actually claim to show that it is false. They contend, for example, that materialism has fought *psi* research tooth and nail because, if proven to be true, it would be fatal to its "ideological system." (Remember Daryl Bem?) They add that an important tenet of materialism is that materialist ideology trumps evidence.[30]

But Beauregard and O'Leary clearly conflate methodological and metaphysical claims here.[31] As we saw in chapter 2, methodological naturalism—also known as science—makes no metaphysical claims and is therefore completely open to the possibility of *psi* or other paranormal phenomena. If this weren't so, Daryl Bem would not have been able to conduct his experimental research on *psi* in the first place! What leads the majority of scientists to reject Bem's claims is not the "ideological" system of materialism but the distinction between good and bad evidence, illustrated by the story of Facilitated Communication. Materialistic science isn't "ideologically" opposed to the reality of Douglas Biklen's technique, either. It's just that when properly controlled experiments and replications are conducted, Facilitated Communication, like *psi*, evaporates.

Nevertheless, Beauregard and O'Leary insist that materialism is deeply flawed because the mind is clearly independent from matter. How do they know this? Thanks to benevolent "nonmaterialist" neuropsychiatrists who treat certain disorders by having patients "reprogram their brains," which Beauregard and O'Leary claim demonstrates that the minds of these patients can change their brains.[32] But this is just circular logic. Step one: presuppose that the mind is separate from the brain. Step two: observe that the mind can change the brain. Step three: triumphantly conclude that the mind must be separate from the brain. This remarkable piece of flawed logic is repeated almost verbatim, with reference to Beauregard and O'Leary, by New Dualist Dinesh D'Souza in his book *Life after Death: The Evidence*.[33] As for Beauregard and O'Leary, the spirited duo unveiled an extremely ambitious agenda in their book. They want nothing less than the replacement of materialist science by "nonmaterialist science."[34] But what exactly is nonmaterialist science? Material science, understood without metaphysical baggage, is simply science. So what exactly are Beauregard and O'Leary proposing to replace science with? Nonscience? Magic?

Another strategy to deal with the evidence supporting materialism is to rely on an intuition pump (a kind of analogy) I mentioned earlier—the radio-brain hypothesis. Proponents of this approach argue that brain states are merely *correlated* with mind states, but that the former do not actually *cause* the latter. Indeed, correlation does not entail causation. My increasing age is positively correlated with our ballooning national debt, but nobody would want to argue (I hope) that my getting older is what is actually causing our economic woes. Enter the radio theory of the brain—the idea that the brain does not cause the mind, but that it merely serves as a gateway for it, just like a radio set functions as a receiver and decoder of electromagnetic waves. In his book, Dinesh D'Souza follows this route and asks how we can be confident that "the brain is a manufacturing plant for the mind and not merely a gateway or a transmission belt."[35]

Indeed, how do we know that our brains are not the unwitting receptors of a cosmic soul signal? If they were, this would seem to leave the door open for the possibility of life after death—precisely what D'Souza claims to demonstrate in his book. A radio set (or a TV, if you prefer) and the signal it receives are separate things, and so they can exist independently of each other. (Remember my analogy for mind-body detachability in chapter 2?) Destroy the receptor, and you still have the signal. Obliterate the brain and you still have the soul. I agree that there is something seductive about this analogy. However, a little thinking tells us that it just doesn't cut the mustard. Philosopher Daniel Dennett calls intuition pumps that backfire "boom crutches" because, although such devices are initially intended to help support an argument—the "crutch" part—they end up blowing up in your face—the "boom" part. The key here again is the notion of consilience.

The manufacturing-plant view of the brain, to use D'Souza's analogy, wins the day because it is supported by everything we know about science. The brain-as-receptor-of-cosmic-soul-signal hypothesis might enjoy some initial plausibility, but a few moment's reflection reveal so many dead-ends, contradictions, and nonsensical implications that it will make your head spin.

For starters, the receptor view of the brain doesn't even begin to respond to the challenge posed for dualism by what we called the fragility of the mind. If damaging only parts of the brain can annihilate just about every aspect of our mind, then by what miracle would the complete destruction of our brain following death leave us with all our mental faculties intact so that we can recognize Uncle Fred in heaven? If the soul needs a functioning brain to be able to think, see, and feel, then how could it perform these functions without a brain at all? And if it could, then why do we need a body and a brain in the first place? Why aren't we just blobs of ethereal soul stuff? At least we wouldn't have to worry about back pains, hemorrhoids, and lung cancer.

Note that this problem isn't new. Thomas Aquinas (1225–1274), who was deeply influenced by Aristotle's conception of the soul, already struggled with it. Without a functioning body, Aquinas believed that the soul would be deprived of all its senses. Forget about recognizing Uncle Fred in heaven, then. Aquinas speculated that the only powers left to the soul after the destruction of the body were those that did not depend on the body in the first place—our faculty of reason and understanding. Had he known what we know today, namely that the faculties of the rational soul depend on the brain, it is not hard to imagine what Aquinas would have concluded.[36]

If our brains function as receptors of soul waves, then why are we not all massively schizophrenic or all carbon copies of each other? After all, the human brain works according to a fixed set of neurophysiological principles. If thirty million Americans can tune in to watch the Super Bowl, what would prevent thirty million brains from receiving the same soul signal or one brain from receiving thirty million signals? And how exactly do soul signals get connected to the right brains? Did my own soul signal have to sit there patiently waiting for billions of years of evolution so that on the fifteenth day of the year 1970 it could finally hook up with my body and generate the thoughts that led to the writing of a book that denies the existence of the soul? Without a signal, radio receivers and TV sets are totally useless. Either you get a signal and you can enjoy the full spectrum of channels and programs, or you don't get a signal and you might as well use your TV set as a paperweight or a fish tank. (If you own a flat screen TV, I would worry about the fish.)

Here's another troubling question: Does the all-or-nothing radio-brain view entail that the soul signal gives rise to my entire mental life? Are the languages I speak, the memories I have, the skills I possess all the product of something beamed into my brain from above? My suspicion is that the reason I speak French and English is because I grew up in France and then moved to the United States. I am also convinced that my memories have to do with the people I've met and the places I've visited in this world. If certain aspects of my mind are the obvious consequence of my dealings with the denizens of the physical world, then what exactly is the soul signal supposed to do? Does it just make me conscious? But conscious of what exactly? The reason I can see the computer monitor sitting in from of me as I type this sentence is that I have a functioning pair of eyes attached to a functioning brain, itself attached to a functioning body. My cat can see the computer too, and she loves attacking my monitor with her paws to try to catch my virtual mouse. Does that mean that my cat has a soul too? Descartes certainly didn't think so. Perhaps, a dualist would retort, my soul makes me self-conscious. But what about people suffering from Cotard's Syndrome? Well, our imaginary dualist could respond, a bad brain can distort the soul signal. But if so, we are back to square one: souls need functioning brains, so how could they function without brains?

Analogies are a useful tool to explain ideas that we actually do understand. They are often used as convenient shortcuts to distill the gist of a more complicated set of ideas and create the desired "ah ha!" moment in our minds. If you were to ask an engineer how a TV works, you would get a very detailed explanation that transcends simple analogies. A physicist would also show you all the relevant equations. Science does understand electromagnetic signals very well; so well, in fact, that we have radio sets, TVs, and iPhones. When we do not understand something, however, analogies have exactly the opposite effect. They serve only to confuse people and give our ignorance an air of credibility. New Dualists like Dinesh D'Souza can talk about receptors and conveyor belts all they want, this should not prevent us from appreciating a simple truth: they do not have the slightest inkling what the soul signal is, how it functions, how it connects with brains, or what it allows brains to do. The analogies they deploy are simply a last-resort trick to try to salvage their doctrine in the face of astronomical odds.

Consilience does tell us that brain activity and mental states are causally linked to one another and not merely correlated with each other. When your dentist gives you a shot of novocaine, she is not relying on a mere correlation between brain states and mental states in hopes that the chemicals will induce the desired numbness in your jaw so you won't bite off her fingers when the drilling begins. Indeed, the injection is what *causes* your conscious perception of pain to temporally fade. Besides, notice that the "it's-only-a-correlation" argument can be applied to just about anything under the sun— which goes to show that such claims are completely toothless (to continue with my dentistry analogy).

You thought the flames you ignited by turning on the burner under your kettle were actually causing the water to boil after a few minutes? Well, don't be so sure. How do you know that the action of the fire and the boiling water are not merely correlated, and that the real job isn't done by an immaterial substance with causal oomph called phlogiston? Ask yourself the same question about your TV and its remote control, your car and how you start it, your smartphone and the virtual buttons you push to call your friend in California, your computer and its keyboard, that egg you dropped and its breaking, the sentences you are now reading and the thoughts that are entering your mind, and you will see the radio analogy of the brain for what it really is: a cheap trick that allows people to confidently talk about something of which they do not possess the slightest understanding.

In *L'Homme Machine*, La Mettrie did more than argue against the existence of the soul. He reflected on the consequences of the demise of dualism, too. If the soul is the mechanism that gives rise to voluntary action, then what does its nonexistence entail? Are we at risk of losing our cherished free will? And without free will, how can we hold people morally responsible for their actions? If the soul is indeed a fiction, why is it such a powerful and widespread one? Why have most people, at most times, and in most places believed that human beings have some kind of soul? These are the questions we will address in the chapters to follow. New Dualists like Dinesh D'Souza believe that the demise of the soul would suck all the meaning out of life.[37] How sad, unimaginative, and untrue. In the middle of the eighteenth century, La Mettrie was ebulliently optimistic about the consequences of materialism, and in the twenty-first century, as we will discover in the chapters ahead, we have even more reasons to rejoice.

Chapter 7

DESCARTES'S SHADOW

Resistance to certain scientific ideas derives in large part from assumptions and biases that can be demonstrated experimentally in young children and that may persist into adulthood.

—**Paul Bloom and Deena Weisberg,**
Childhood Origins of Adult Resistance to Science, 2007

We owe one of the most trenchant critiques of Cartesian dualism ever penned to the philosopher Gilbert Ryle (1900–1976). In his classic 1949 book, *The Concept of Mind*, Ryle argued that substance dualism is not merely false, it is one big mistake—what he called a *category-mistake.* To illustrate this kind of error, Ryle asked us to think about a foreigner visiting Oxford University. After being shown all the buildings, the playing fields, the libraries, the academic departments, and the administrative offices, Ryle's inquisitive visitor would ask: "But where is the university?" The mistake, of course, is to assume that "the university" refers to another building on campus in the same way that "the art museum" or "the Physics Department" does. Terms like "the university" and "the library" belong to different categories. For Ryle, treating the mind as a separate substance, the way Descartes did, was akin to counting "the university" as another building. In his book, Ryle ridiculed the Cartesian doctrine, famously calling it the dogma of "the ghost in the machine."

But if substance dualism is such a terrible idea, why is it so widespread? Why have most people, at most times, and in most places believed in the existence of souls? Some have argued that this is because ordinary folks are poor thinkers compared to professional philosophers and scientists. Or maybe it is because soul beliefs stem from a natural yearning for comfort in the face of death, or a general human need to explain puzzling phenomena. In his wonderful book *Religion Explained*, anthropologist Pascal Boyer briefly considers these commonplace answers but quickly dismisses them. I think Boyer is right in being doubtful that poor thinking skills or a general tendency to be superstitious can shed much light on the origins of soul beliefs.

The near universality of those beliefs suggests that they may come to us very naturally. This is what a number of cognitive scientists interested in the origins of supernatural concepts have begun to demonstrate. Drawing on these recent developments, I will suggest that the origin of soul beliefs can be better understood if we pay closer attention to the basic machinery of the human mind. There is now solid evidence that the relevant psychological principles are universal across the species and that they emerge early in the lives of human beings. To the extent that dualism arises as a natural consequence of the operation of these core psychological principles, we may have a way to account for the ubiquity of soul beliefs. If human beings are natural-born dualists, as psychologist Paul Bloom has argued, then we all live in Descartes's shadow.

The origins of soul beliefs would probably have remained a mystery in the view of human psychology espoused by Ryle and his contemporaries. During the first half of the twentieth century, North American psychology was dominated by a school of thought called behaviorism. Viewed from that perspective, "mentalism" reeked of subjectivity and had to be banished from the study of psychology if the discipline had any hope of achieving the status of a real science. Mental concepts such as beliefs, goals, intentions, memories, and the like were replaced by more concrete notions like stimuli (think of a ringing bell) and responses (think of a dog salivating) that could be observed and measured objectively.

Not long after the publication of Ryle's book, however, behaviorism began to lose steam and was soon eclipsed by the cognitive revolution of the 1950s. With the demise of behaviorism, mentalism made a spectacular comeback, and today the mind has reclaimed center stage in the study of psychology. Descartes's shadow still looms large. For people watching from the sidelines, this state of affairs can be puzzling. If the mind is taken seriously again, but it isn't a separate substance in Descartes's sense, then what is it? And if mental processes eventually reduce to physical operations in the brain, as we saw in the previous chapter, then why bother talking about the mind at all? The answer to these questions, I will show you, can be found in Ryle's analogy. The study of minds and brains does not entail the existence of two different substances, but rather the identification of two different and complementary perspectives on the same problem.

The reason Descartes postulated his second substance, *res cogitans*, is that he could not doubt that he was a thinking being, and he also could not imagine how the physical body can give rise to something as ineffable as conscious thought. Today, consciousness remains a mystery wrapped inside an enigma. In his book *How the Mind Works*, Steven Pinker humbly admits that consciousness "beats the heck" out of him. More than three centuries after Descartes's famous *cogito ergo sum*, the problem of consciousness still bamboozles philosophers and neuroscientists alike. The dark veil cast by Descartes's shadow remains impenetrable. Nevertheless, some of the New Dualists we met in earlier chapters gleefully turn our ignorance on its head and declare that consciousness proves the existence of the soul. Following Feynman's first principle, I will show you that there is no reason to fall for that trick.

The demise of dualism we chronicled in the last three chapters gives rise to questions about the origins of soul beliefs, the nature of mind, and the problem of consciousness. These are the scientific questions we will touch on in this chapter. For many people outside of academia, the idea of a world without souls also produces dizzying existential questions. By letting go of the soul, have we opened Pandora's box? Have we condemned ourselves, like Harry and Rodrick in *The God Gun*, to look into each other's eyes and find that everything has lost its meaning, purpose, and beauty? I have reserved the final two chapters of the book for an in-depth treatment of these questions. For the time being, let us take one last stroll through the hallways of the ivory tower before we make our final steps into the soul-free world.

THE ORIGINS OF SOUL BELIEFS

In an essay published shortly after the outbreak of World War I, Sigmund Freud wrote that our own death is unimaginable to us and that whenever we try to think about it "we can perceive that we really survive as spectators."[1] For Freud, soul beliefs emerged as a solution to what the philosopher Steven Cave calls the Mortality Paradox. We know that we are going to die, and yet being dead cannot be consciously simulated, and so it is unimaginable to us. The doctrine of the soul resolves the Mortality Paradox by negating its second premise. The body eventually dies, but the soul lives on, and so consciousness is never extinguished. There may be a grain of truth to Freud's approach, and we will come back to the Mortality Paradox in the next chapter. Nevertheless, fear of death cannot be the whole story. After all, there are many other comforting illusions that we could entertain in the face of death. So why the soul doctrine and not something else?

A more promising approach comes from recent work in cognitive science on the origins of supernatural concepts. This exciting new line of work reveals that magical concepts like the immaterial soul are not that different from the more mundane concepts we routinely entertain. To the extent that ordinary cognition imposes constraints on the kind of supernatural concepts we are likely to entertain, we may discover there is only a limited and predictable catalogue of such concepts. If so, we may be able to answer the question left open by Freud's musings: Why souls and not something else? Here are five ideas, some drawn from the new work on the origins of supernatural concepts, that I believe can take us beyond the Freudian analysis, on the path to a deeper understanding of the origins of soul beliefs.

- Our intuitive psychology is inherently dualistic. According to psychologist Paul Bloom, we are natural-born dualists.
- Explicit dualistic beliefs emerge early in development and can already be found in preschool children. Young children also believe that the mind, unlike the body, can survive death.
- There are good reasons, based on everyday experience, to treat the conscious mind as separate from the body.
- Supernatural concepts like soul beliefs are minimally counterintuitive concepts. This helps understand their characteristics and cultural prevalence.
- The United States is an unusually religious country in which soul beliefs are constantly reinforced.

A good starting point to begin thinking about our intuitive psychology, the first idea on my list, is the Terminator analogy I introduced in chapter 2. If we were to design a robot like the Terminator, capable of intelligently navigating our world, we would need to make sure that it was programmed with the right expectations about its surroundings and the entities it would be likely to encounter. For one thing, our robot would need to understand the basic physics governing its environment. It would need to know that material objects are solid, that they are subject to the principle of gravity, that we can interact with them through physical contact, and so on. Such knowledge would allow the robot to navigate its environment efficiently, without getting stuck trying to walk through walls or risking self-termination by falling off cliffs or rooftops. Along with a "physics module," we would also need to program our robot with a "psychology module," especially if its ultimate goal is to terminate Sarah Connor. Our killer robot wouldn't be very efficient if it merely expected its human target to be a physical object that is solid and obeys the principle of gravity. Unless the Terminator also expects Sarah to be driven by *goals* (escaping its deadly grip) and *intentions* (staying alive), and to be able to act accordingly, our robot's chances of successfully accomplishing its mission would be pretty slim.

Human beings, too, live in a world populated by objects and agents. Our goal in life isn't to terminate Sarah Connor, of course, but survival ranks pretty high on the list. Being able to distinguish a pile of leaves from a hungry predator would have made a big difference for the survival of our ancestors. We might therefore expect that biological evolution, the equivalent of the "programmer" in my Terminator analogy, would have endowed us with a built-in ability to reason about objects and agents. Over the past several decades, this hypothesis has been confirmed by a wealth of findings from developmental psychology. Research on human infants, children, and adults across different cultures indeed provides compelling evidence for the existence of two separate psychological systems dedicated to reasoning about objects and agents—"intuitive physics" and "intuitive psychology."

Human babies are "intuitive physicists" in the sense that they expect material objects to be solid and cohesive, to trace continuous spatiotemporal paths when they move, to fall to the ground when unsupported, and to not spontaneously move unless they are acted upon. In one study, psychologist Elizabeth Spelke showed that four-month-old infants are sensitive to the principle of spatiotemporal continuity.[2] Spelke and her team relied on an ingenious experimental procedure that involves keeping track of how long infants look at one object or event. If infants are presented with the same event repeatedly, they will quickly get bored and start looking away. When presented with subsequent events, they will tend to look longer at the ones they perceive to be novel compared to their earlier experience. This preference can be observed with infants as young as one day of age![3]

In their experiment, Spelke and her team presented infants with events in which a thin object moves across a stage and passes behind two occluding screens, separated by a small distance. In the continuous version of the event, the object starts in full view, moves behind the first screen, reappears to traverse the space between the two screens, passes behind the second screen, and finally reemerges to the right of that screen (see figure 7.1). In this version, the motion of the object obeyed the continuity principle because it traced a continuous spatiotemporal path as it moved behind the two screens. The discontinuous version of the event was similar to the continuous one except for one crucial difference: after the object disappeared from view behind the first screen, it magically reappeared to the right of the second one *without visibly traversing the space between the two screens.*

In a second phase of the experiment, infants were presented with events displaying the motion of either one or two separate objects (see bottom part of figure 7.1). If infants are sensitive to the principle of continuity, they should treat the continuous-motion events as involving only one object moving in and out of view. During the testing phase, they should look longer at the display involving two separate objects (two is novel compared to one). By contrast, infants should infer the presence of two objects in the discontinuous events, because a single object cannot disappear from one location and reappear at a different one. During the testing phase, infants should therefore look longer at displays involving only one object (one is novel compared to two). This is exactly what Spelke and her team found, showing that infants can reason about physical objects using the principle of continuity.

One of the most compelling demonstrations that human beings have different modes of construal for objects and people—intuitive physics and intuitive psychology—is to show that the principles that operate in one domain do not apply to the other. Building on Spelke's work, a team of psychologists at Yale University led by Karin Wynn and Paul Bloom investigated the question of whether five-month-old infants would recognize that people are material objects too, also subject to the principle of continuity.[4] After all, people have bodies, and bodies are physical things. In order to find out, the Yale team ran the experiment I just described, but with an important twist. They presented participants with short video clips featuring either an inanimate block gently gliding across a stage and passing behind two occluding screens or a human female walking across the stage and passing behind the occluders. Remarkably, the Yale team found that infants expect material objects to follow the continuity principle, replicating Spelke's earlier finding, but not people. In other words, infants do not seem to initially regard people as material objects!

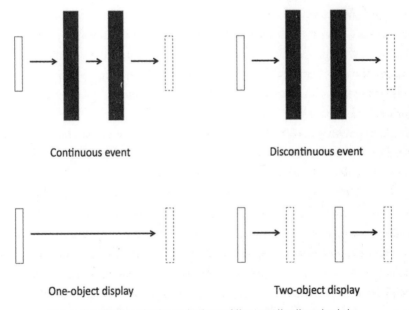

Continuous event   Discontinuous event

One-object display   Two-object display

Figure 7.1: Testing for knowledge of the continuity principle.

This doesn't mean that babies think that people are ghosts. We all come to realize that our bodies, just like other material objects, are subject to the laws of physics. However, this intriguing result suggests that there is something special about people, that we regard human beings as different from mere physical objects. Our intuitive psychology tells us that people, unlike billiard balls, are self-propelled agents. Nobody needs to exert physical contact on my body for me to spontaneously stand up from my chair, walk to the kitchen, and pour myself a glass of water. Other human beings can make sense of my behavior by assuming that I am a creature driven by *beliefs*, *goals*, *desires*, and *intentions*—the stuff of minds. This is the fundamental difference between the two systems. Our intuitive physics tells us that the world contains material objects whose behavior is governed by the principles of solidity, contact, cohesion, and continuity. Our intuitive psychology tells us that agents have invisible minds that control their behavior through forces like goals, beliefs, and intentions.

We just saw that infants initially apply the principles of intuitive physics to material objects but not to people. The reverse is also true. Infants apply the principles of intuitive psychology to people but not to inanimate objects. In a series of elegant experiments using the visual-habituation paradigm, psychologist Amanda Woodward showed that infants regard the actions of people, but not those of inanimate objects, as goal-driven. In one experiment, Woodward habituated infants to scenes in which a human actor reached for and grasped one of two toys sitting next to each other on a stage.[5] Next, the location of the toys was switched, and infants witnessed events in which the actor reached for the target or the nontarget toy. Infants looked longer when the actor grasped the nontarget toy, suggesting that they are sensitive to the actor's goal (reaching for one particular toy). Importantly, this effect did not obtain when the actor's arm was replaced by a mechanical claw that moved to grasp the toy. For infants, only the actions of people are driven by goals.

Infants can also reason about other aspects of people's minds such as their beliefs. Imagine a situation in which a child has eaten the last cookies in a box, unbeknownst to her brother, and then later on sees him reach into the box. To make sense of her brother's behavior, the child must understand that his actions are not based on the way the world *is* but rather on what her brother *believes* it to be. The child must take into account her brother's *mental state* and understand that he inaccurately believes there are still cookies in the box. The developmental psychologist Renée Baillargeon and her team demonstrated that fifteen-month-old infants correctly reason about other people's beliefs, even if those beliefs are false.[6] In the experiment, a human actor placed a small toy inside one of two boxes in full view of the infants. The actor then left the scene and returned later to look for her toy. During the actor's absence, the infants saw that the toy was either moved into the other box or remained in the original one. Where would the actor look for her toy when she came back? The infants were no fools. They always expected the actor to look for her toy where she *believed* it to be, regardless of the truth of those beliefs.

All this shows that human beings have separate modes of interpretation for objects and for people—intuitive physics and intuitive psychology. When applied to people, these two systems tell us that we are material bodies whose behavior is controlled by invisible minds. Or, alternatively, that we are invisible minds who affect the physical world through the actions of our material bodies. Looking at human nature through the prism of cognitive psychology, Paul Bloom concludes that we are natural-born dualists who see the world as being composed of bodies and souls.

The instinctive distinction between material bodies and immaterial minds provides a natural basis for the development of soul beliefs. However, the gap between intuitive dualism and explicit soul beliefs needs to be bridged. The second idea from my list is that this gap is bridged early in our lives. Young children in their preschool years already have an explicit understanding that minds and bodies are not cut from the same cloth. In their book *Children Talk about the Mind*, psychologists Karen Bartsch and Henry Wellman present evidence supporting this conclusion. They show that preschoolers already understand that mental entities like thoughts, dreams, and memories are not real in the same way that material objects like cups, chairs, and tables are. Preschoolers also know that mental states are inherently private rather than public, which makes them very different from material objects like their toys. Importantly, dualistic beliefs in preschoolers, unlike those of preverbal infants, manifest themselves *explicitly*, in what these children say and the way they answer questions.

A fascinating illustration of this phenomenon comes from a study by psychologists Jesse Bering and David Bjorklund, who told young children the story of a mouse who was killed by an alligator: "Well, it looks like brown mouse got eaten by Mr. Alligator. Brown mouse isn't alive anymore." Children were then asked a series of questions about the dead mouse. Some were biological and had to do with the mouse's body: Will the mouse ever need to *eat* food again? Will he ever need to *drink* water again? Will he *grow up* to become an old mouse? Other questions were about the mouse's mind: Does the mouse *know* that he's not alive? Does he still *want* to go home? Does he still *love* his mom? Most children between the ages of three and six understand that death implies the cessation of bodily activities and explained to the experimenters that the mouse won't need to eat food or drink water again, and that it won't grow up to become an old mouse. Strikingly, however, the same children also believed that mental activity would continue after death. They almost all said that the dead mouse knew he was not alive, wanted to go home, and still loved his mom.[7]

What could lead us to such explicit dualistic beliefs? The third idea from my list is that there are good reasons to believe, based on everyday experience, that the physical and the mental are two separate domains. A ubiquitous source of explicit dualism comes from the way we talk about ourselves. I can tell people about "my hand," "my foot," or "my brain" as though the real me was a separate "I" or "self" to whom my body belongs. The radical difference between body and mind is also something that we all experience routinely in the private theater of our own consciousness. Right now, if I want to grab the cup sitting in front of me on my desk, all I have to do is consciously relay my desire to my right arm and this will immediately be followed by the appropriate action. From the inside, consciousness is perceived as a powerful causal force that we use to control our body. The trick, of course, as we saw in chapter 3, is that we are not privy to the intricate neuronal dance that takes place under the hood and gives rise to our conscious thoughts. We therefore perceive consciousness as an uncaused cause, an immaterial genie that magically surfaces when we need it.

Psychologist Daniel Wegner calls this trick "the illusion of conscious will." The illusion for Wegner comes from the conscious feeling of authorship that typically accompanies our voluntary actions. But correlation does not entail causation. Consciousness may not be the actual cause of our actions. To support his view, Wegner points to all the cases where our actions and the accompanying experience of authorship do not go hand in hand. Sometimes we can perform actions without being consciously aware that we are the author of these actions. This is what happens with Facilitated Communication and the Clever Hans effect that we discussed in chapter 3. This is also what happened to Wilder Penfield's patients, as we saw in chapter 6. When their brains were electrically stimulated, the patients performed what appeared from the outside to be voluntary actions. But when asked about these movements, the patients reported no conscious desire to act, and told Penfield that it wasn't them who performed the actions. Conversely, we can also sometimes believe that we performed certain actions whereas in fact nothing of the sort took place. This is exactly what happened when the team of French neuroscientists we met in chapter 6 was able to trick patients into believing that they were the authors of actions that they in fact never performed.

On the view developed so far, the concept of a supernatural soul represents an explicit manifestation of an early-emerging form of intuitive dualism. Early in life, this intuitive dichotomy becomes explicit, is reflected in the way we talk, and represents an important aspect of our private experience. The fourth idea from my list is that soul beliefs are *minimally counterintuitive concepts*, which explains why their cultural transmission has been so successful. This hypothesis was proposed and popularized by the anthropologist Pascal Boyer in his book *Religion Explained*. The idea is that supernatural concepts are variations of natural categories like persons, animals, plants, artifacts, and natural objects. They are minimally counterintuitive in the sense that they violate only one or a few expectations for members of a given category while conforming to all the others. For example, the concept of a soul is a minimal variation on the concept of a person. A soul is simply a person who happens to not have a body. This is the minimally counterintuitive part. We otherwise reason about souls in the same way that we reason about people. Souls are conscious, they can feel and remember, they can see us, and we can even talk to them. (At least for those of us who believe that communication with the dead is possible.) What makes minimally intuitive concepts particularly potent from the point of view of cultural transmission is that they are unusual enough to grab our attention—a person without a body, really?!—and yet, they are similar enough to natural categories so as to be easily remembered.

One of the virtues of Boyer's hypothesis is that it can easily be put to the test. In a recent study, Elizabeth Spelke and her team at Harvard asked whether seven- to nine-year-old children who listened to stories involving intuitive, minimally counterintuitive, and maximally counterintuitive concepts would remember the minimally counterintuitive ones better. In one story, children were told about a normal lizard that "ate insects off the ground and crawled around quickly on all four of its feet." In another story, children heard about another lizard "that had a long thin tail and could never die no matter how old it was." This lizard is mildly odd because children know that all animals eventually die. This is our minimally counterintuitive lizard. In a yet another story, children heard about a lizard "that always melted in the hot sun, could never die no matter how old it was, and could hear other creatures' thoughts." This critter is maximally weird! It violates our intuitive sense of physics, biology, and psychology. As predicted, children remembered the stories involving minimally counterintuitive concepts better and in greater detail.[8]

So far, everything I have told you applies equally well to people who grew up in the United States, France, or Germany. So why are there such drastic differences in the rates of soul beliefs in these different parts of the world, as we saw in chapter 2? The final idea from my list is that the United States is an unusually religious country among developed nations. This may not be an obvious conclusion to someone who grew up here in America and has had little exposure to other cultures. However, as anyone who has ever lived abroad or looked at comparative data knows, the United States sticks out like a sore thumb. The number of Americans who report believing in God, religious miracles, heaven, and angels is arresting. A 2008 Pew Global Attitudes Project provides a good measure of the importance of religion as a function of wealth. The United States, the richest nation in the world, has levels of religiosity unparalleled in other wealthy nations and almost comparable to those seen in third-world countries.[9] As linguist and political activist Noam Chomsky once remarked, "You'd have to maybe go to mosques in Iran or do a poll among old ladies in Sicily to get numbers like this. Yet this is the American population."[10]

As a result of the excessive level of religiosity found in the United States, anyone running for public office in America must make it a point to declare his or her religious faith. Atheists are generally despised in our culture and most people would not want to vote for a presidential candidate who does not believe in God.[11] It is not uncommon for preachers to appear on national television and openly declare that, according to the Bible, homosexuality is a sin.[12] Some elected officials have gone so far as to argue that we should not worry about climate change because, according to the book of Genesis, God will be the final arbiter on such matters.[13] The United States is also the only wealthy nation in which a minority of the population believes in evolution. The religious war against Darwin has been raging for decades in America. In such an overtly religious culture, perhaps we shouldn't be too surprised to find that most people believe in the soul and its survival after death. In western European countries, which are also among the wealthiest nations, religiosity levels are a fraction of what they are here in the United States, and belief in the soul is correspondingly much less pronounced, as we saw in chapter 2.

THE RISE OF THE MODERN MIND

I spent the last three chapters trying to convince you that the mind isn't a separate substance and that people do not have immaterial souls. You might now wonder what the mind actually *is* and why we should even bother talking about it anymore if everything mental boils down to physical activity in the brain. From a scientific point of view, the question of what the mind *is* has a rather straightforward answer that ties in with everything else I have told you. In the pages ahead, I will continue my demystification act by showing you that talk about the mind is in fact talk about the brain *at a certain level of abstraction*. I will then show you why invoking a mental level of analysis can be a very productive strategy if one is in the business of trying to understand the behavior of intelligent creatures.

To begin illustrating the idea that the mind is the brain viewed from a different perspective, consider the picture in figure 7.2. Try looking at that picture with your eyes as close to the page as possible and you will see a meaningless assortment of pixels. You could describe these pixels by saying that some of them are white, some black, and others different shades of grey. You could even pick out a single pixel, describe its shape and color, and specify its position with respect to any other pixel in the grid. Next try positioning your book or tablet as far away as you can and look at the picture again. You should now see someone familiar emerge—Elvis Presley. From this new vantage point, the individual pixels have fused into a meaningful image that anyone can instantly recognize.

The trees have disappeared and you can now see the forest. And yet, you are looking at *exactly the same physical array of pixels as before*. If you wanted to describe what you are seeing from this new perspective, you would no longer talk about pixels. Instead, you would be using expressions like *the rock star*, *the King*, or *Elvis Presley*. You could even explain certain facts about your behavior using these expressions, facts that you would have trouble explaining talking about pixels. If someone asked you why you were able to identify the man in that picture so quickly, you could say that it is because Elvis is a cultural icon, that he is instantaneously recognizable, and so on. It would be much harder, if not impossible, to answer the same question if all you could talk about were pixels, their color, and their position in the grid.

This simple example illustrates the idea that we can talk about the same physical object from different perspectives, or at different levels of analysis.

When cognitive scientists talk about the mind, they are really talking about the brain at a different and more abstract level of analysis. Taking a mentalistic perspective to talk about the brain, as we will discover, has important scientific virtues and naturally ties together several key notions that I introduced in earlier pages. As Noam Chomsky, one of the fathers of the modern mind, wrote, "Talk about the mind [is] talk about the brain at an abstract level at which, so we try to demonstrate, principles can be formulated that enter into successful and insightful explanation of . . . phenomena that are provided by observation and experiment. . . . Mentalism, in this sense," Chomsky continues, "has no taint of mysticism and carries no dubious ontological burden."[14]

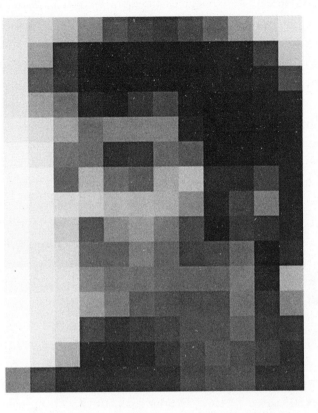

Figure 7.2: Pixelated illusion. Image from *PsychWorld*.

Let me now turn to another simple example to show you how mentalism can be used to predict people's behavior. Suppose we go back to the 2012 presidential election—say to July of that year. Imagine that we were interested in predicting the behavior of the two candidates, Barack Obama and Mitt Romney. To make things more concrete, suppose we wanted to predict where the two men would be on the evening of October 3, three months into the future from our perspective. Obama and Romney traveled a lot during the campaign, so they could be just about anywhere. We could think of the candidates as material beings subject to the laws of physics and try to predict their future locations the way we predict the future positions of celestial bodies. We would take into account the previous history of the candidates, the sum of all the forces impinging on them up to the point of our prediction, and from that we would try to calculate their future locations.

It doesn't take a trained physicist to see that this approach would be hopeless. It would be like trying to explain why a segment from the *Colbert Report* is funny by looking at the pixels on your TV screen instead of focusing on higher-level abstractions like current political situations, presidential campaigns, hypocrisy, or plays on words. If we want to predict the way people are going to behave, at least when voluntary actions are involved, we cannot simply rely on the *physical stance* (treating people as billiard balls). We need to focus on a different level of analysis and adopt what Daniel Dennett has called the *intentional stance*. This involves viewing people as rational agents endowed with minds driven by goals, desires, and beliefs. From this new perspective, it becomes much easier to predict the future locations of the candidates, at least some of the time, even months in advance. If we knew that Obama and Romney *believed* that there would be a presidential debate on October 3, 2012, at the University of Denver, and if we also knew that both candidates *wanted* to participate in that debate, then we would be able to correctly predict, barring any unforeseen disasters, that on the evening of October 3, 2012, the candidates would both be at the University of Denver.

The weather cannot be predicted more than two weeks in advance, and people are at least as complex as cloud formations, so being able to predict the exact locations of the two candidates months in advance is no small feat. Notice that the intentional stance is purely methodological and only treats mental entities as explanatory constructs. Beliefs ultimately reduce to physical operations in the brain, the way the picture of Elvis ultimately reduces to a series of pixels that bear no resemblance to the rock star. On this view, mentalism is nothing more than methodological naturalism applied to psychology. Because mentalism is a mode of explanation that carries no suspicious metaphysical baggage, it can naturally be extended to other creatures, including human infants, nonhuman animals, and even intelligent machines, as we will see shortly.

The experiments on infants we reviewed earlier in this chapter were all motivated by mentalistic assumptions. The logic goes something like this: if we assume that at the mental level infants know principle X, we can predict that they will do Y and Z in the carefully controlled experiments that we designed. To the extent that these predictions are borne out, the mental construct "principle X" plays the role of an explanatory concept via which researchers can make sense of the behavior of the infants. This is what we saw with Spelke's principle of continuity, Baillergeon's conclusion that infants can reason about the beliefs of other agents, and Woodward's finding that infants ascribe goals to agents but not to inanimate objects.

Applying the intentional stance is something that we all do very naturally. In 1944, the psychologists Fritz Heider and Mary-Ann Simmel created a short movie in which simple geometric shapes—two triangles and a circle—interacted with each other in ways that were purposefully anthropomorphized. The large triangle came out of an enclosed space and started chasing the small triangle and then the circle. The small triangle and the circle then teamed up to deal with the aggressive big triangle and even managed to trick it back into the enclosed space, seemingly rejoicing over their accomplishment. When adult participants were shown the movie, they couldn't help but describe the geometric shapes as though they had minds and were driven by goals and intentions. The big triangle was a bully who *wanted* to hurt the other triangle, the small triangle *tried* to escape, and the circle was a hero who *helped* the small triangle.

We also naturally adopt the intentional stance when we try to understand the behavior of intelligent machines. If you were playing chess against a computer, and hoped to not be embarrassed too quickly, you would need to predict the computer's chess-playing behavior. How could you do that? A very useful approach would be to adopt the intentional stance and assume that the computer *knows* the rules of chess, *believes* that your moves are intended to defeat it, and acts as though its ultimate *goal* was to checkmate you. This would be much easier—and faster—than poring over endless strings of 1s and 0s or trying to deduce the computer's next move by considering the electrical currents passing through its circuits. If you threatened the computer's queen with one of your knights, and saw the computer evade your attack, you would naturally conclude that the computer *believed* that you were trying to capture its queen. If the situation were reversed, you would think that the computer *wanted* to capture your queen.

What is particularly interesting about the computer analogy is that it reveals the intentional stance for what it actually is—a useful *methodological* approach. In acting and talking as though computers were rational agents driven by goals, desires, and beliefs, we certainly do not imply that they possess a mysterious incorporeal mind. The reason we can be so sure is that we build computers, and so we know that there are no mysterious spirits, vapors, or souls inside of them! Descartes might have been tempted to conclude that the GPS device in my car contains an immaterial soul because it seems to know where I am, can reason about how to reach my intended destination, and can even intelligently talk to me about it. And yet we know for a fact that GPS systems are pure *res extensa* without an ounce of *res cogitans* inside. Adopting the intentional stance and talking about the minds of infants, nonhuman animals, and machines is simply a useful methodological choice. Only by taking the additional metaphysical step of insisting that the mind must be a separate substance, as opposed to a mere abstraction, do we commit Ryle's category mistake.

THE MYSTERY OF CONSCIOUSNESS

At the dawn of human civilization, the Chinese philosopher Confucius suggested that real knowledge is to know the extent of one's ignorance. Reflecting on the foundations of human knowledge, the linguist Noam Chomsky proposed that our ignorance can be divided into two categories: *problems* and *mysteries*. When we face a problem, we may not know how to solve it, but at least we have a sense of what the answer should look like. But when we face a mystery, "we can only stare in wonder and bewilderment," Chomsky writes, "not knowing what an explanation would even look like."[15] For most of human history, the mind was a complete mystery, as we saw in chapter 2. During the second half of the twentieth century, many of these mysteries were upgraded to problems. This is what prompted Steven Pinker to write his classic book *How the Mind Works*. Nevertheless, there remain a number of mysteries about the mind, and none appear to be more baffling than the mystery of consciousness.

If you talk to people about "intelligent" computers that can "think," and describe them as though they have "minds," one of the first things you'll hear back is that there sure is one thing that computers, and machines more generally, cannot do. They cannot *feel*. It's not that machines could not be programmed to mimic the outward manifestations of feelings. It's just that if they did, it would all be a sham. Machines could *act* as though they felt, but they wouldn't really be *experiencing* the private, subjective sensations that define the human condition—or so most people believe. To paraphrase philosopher Thomas Nagel, there wouldn't be anything *it is like* to be a machine (unless you're an anthropomorphized robot like the Terminator). This aspect of consciousness, the inherently subjective nature of experience, is what philosophers call *qualia*. Understanding how physical events in the nervous system can give rise to *subjective* experience is such a baffling problem that it has its own name: The philosopher David Chalmers called it the *hard* problem. In comparison, all the other problems about human intelligence are called the *easy* problems. There is, of course, nothing trivial about these problems, and it might take cognitive scientists another century or two to solve them. They are called "easy" because they are *problems*, not *mysteries*. The hard problem is a genuine *mystery*.

Here's a classic thought experiment proposed by Australian philosopher Frank Jackson, designed to pump our intuitions about the hard problem: Mary is a brilliant scientist who has been investigating the world from a black-and-white room and via a black-and-white television screen. During her entire life, all she has ever experienced are different shades of grey. Mary specializes in the study of vision, and in the course of her investigations, she has acquired *all the physical information there is to know* about the neurophysiology of color vision. She has learned about what goes on in people's brains when they see ripe tomatoes or the sky, and she also knows what happens when people use terms like *red* or *blue*. Mary has also learned about all the detailed effects of various wavelength combinations, how they stimulate the retina, and how the information is subsequently processed by the nervous system. Jackson then asks us to imagine what will happen when Mary is released from her black-and-white room and can finally experience the world of colors. Will she learn anything new? According to Jackson, it is obvious that she will. If so, it follows that her previous knowledge was *incomplete*, in spite of the fact that she had *all* the relevant physical information. *Ergo* qualia are not physically reducible.

This, in a nutshell, is the problem of phenomenal experience, or qualia. What are we to make of it? Among philosophers and neuroscientists, many believe that the problem is real. Some are hopeful that it will one day be solved by science, whereas others—a group that philosopher Owen Flanagan dubbed the *mysterians*—believe that qualia will forever remain out of our reach. If the mysterians are correct, then the hard problem of consciousness is like quantum mechanics to cats or plate tectonics to goldfish. Since we are organic creatures and not angels, as Noam Chomsky is fond of reminding us, our cognitive capacity must have a scope and limitations (like that of other animals). The problem of consciousness may therefore very well fall outside the scope of our science formation capacity, just like quantum mechanics and plate tectonics fall outside the cognitive capacity of cats and goldfish.

There are others, like philosopher Daniel Dennett and cognitive neuroscientist Stanislas Dehaene, who believe that the problem of qualia is a false problem. In his book *Intuition Pumps and Other Tools for Thinking*, Dennett argues that Jackson's thought experiment is nothing but a boom crutch. (We talked about those in the previous chapter.) Here's a different ending to Mary's story that Dennett encourages us to consider. I'll let you judge for yourself whether it successfully deflates Jackson's intuition pump:

And so, one day, Mary's captors decided it was time for her to see colors. As a trick, they prepared a bright blue banana to present as her first color experience ever. Mary took one look at it and said "Hey! You tried to trick me! Bananas are yellow, but this one is blue!" Her captors were dumbfounded. How did she do it? "Simple" she replied. "You have to remember that I know *everything*—absolutely everything—that could ever be known about the physical causes and effects of color vision. So of course, before you brought the banana in, I had already written down, in exquisite detail, exactly what physical impression a yellow object or a blue object . . . would make on my nervous system. So I already knew exactly what *thoughts* I would have. . . . I was not in the slightest surprised by my experience of blue. . . . I realize it is *hard for you to imagine* that I could know so much about my reactive dispositions that the way blue affected me came as no surprise. Of course it's hard for you to imagine. It's hard for anyone to imagine the consequences of someone knowing absolutely everything physical about anything!"[16]

For hard-problem skeptics, it is simply too early to tell whether there actually *is* a hard problem, a problem that science cannot solve. After all, many mysteries were once deemed unsolvable—until they were solved. In 1895, Lord Kelvin confidently declared that heavier-than-air flight was impossible, only to be proven wrong by the Wright brothers a few years later. The idea that we may one day send an object into space was once deemed preposterous, until *Sputnik* was launched into orbit in 1957. In 1934, Albert Einstein claimed that nuclear energy would never be attainable because it would require that atoms be shattered at will. The same year physicist Enrico Fermi proved Einstein wrong. The list goes on. As for the hard problem of consciousness, skeptics believe that we first need to solve all the "easy" problems, which is going to take awhile anyway, and only then, if there is still a mysterious residue when we are done, may we begin to think about reconsidering our basic assumptions. Until then, we should wait and see how far we can get by doing basic science.

This is the point at which I'd be tempted to tell you, "Look, we just don't know how consciousness works." The best we can do, following Confucius, is to admit our own ignorance. But alas, this would be leaving the last word to the New Dualists. Their strategy should be familiar by now: You don't know how the creative aspect of language use works? Then it must be the soul! Science cannot explain qualia? Well, here's more evidence for existence of the soul! The recent books that try to "prove" the existence of the soul all invoke consciousness as a key piece of evidence. Here are some quotes from *Life after Death: The Evidence*, *The Soul Hypothesis*, *The Spiritual Brain*, and *Proof of Heaven*:

- "We have now found two central features of human nature—consciousness and free will—that are irreducible to matter and appear to be independent from it."[17]

- "What gives conscious experience its particular subjective character? How could electrochemical events in the brain . . . produce the taste of a lemon, the distinctive sound of a flute, or the pain of a toothache? (the problem of qualia)."[18]
- "There are good reasons for thinking that the evidence for materialism will actually never arrive. For example, there is the problem of qualia."[19]
- "In fact, the greatest clue to the reality of the spiritual realm is this *profound mystery* of our conscious existence."[20]

Richard Feynman, who gave us the first principle we explored in chapter 3, also remarked that it is much more interesting to live not knowing than to have answers that might be wrong. But for the New Dualists the water of ignorance can be turned into the wine of knowledge and presented as an argument supporting the existence of the soul. The fallacy should now be easy to detect. It is yet another example of the soul-of-the-gaps argument that I warned you about in chapters 3 and 4. The argument fails because it magically takes us from ignorance to knowledge: We are ignorant of the mechanisms responsible for qualia, but somehow we (the New Dualists) know that those mechanisms will never be explained by science (thanks to our powerful crystal ball!), and we also know that the soul must be what is giving rise to qualia (after all, what else could it possibly be?). This all-too-common fallacy was already known to the Greeks as the argument from ignorance. Deriving support for the soul from our ignorance of how consciousness works is what Richard Dawkins would call "a most extraordinary piece of warped logic."

The argument from ignorance is not only logically bankrupt, it also explains nothing. Saying that the soul gives rise to qualia is about as informative as saying that David Copperfield can make his assistant disappear using "magic." Suppose you tell a friend, "Gee, we really don't know how consciousness works . . ." to which she replies, "Our souls make us conscious!" Do you now understand consciousness any better?

The New Dualists all commit the same logical fallacy when they desperately try to squeeze knowledge from ignorance, in the process reaching outlandish conclusions, but Dinesh D'Souza might be considered the most successful at pursuing such conclusions. D'Souza follows the standard superstitious plot when he confidently asserts that consciousness is "irreducible to matter." His second claim, that consciousness appears to be "independent from matter," is even more remarkable. In a parody of former president George W. Bush, the comedian Will Ferrell explains to his audience that liberal scientists are trying to make him look bad by using such things as facts and scientific data. Someone playing the role of an advisor immediately interrupts the comedian turned president to remind him that facts are real and undisputed. Ferrell's brilliant reply: "How do you know that?"

In asserting that consciousness is independent from matter, Dinesh D'Souza, like Will Ferrell in his parody of George W. Bush, is asking us to believe that facts are neither real nor relevant. We may not know *how* consciousness arises from neural computation, but there is little doubt that consciousness *is* intimately related to what goes on in the brain. We saw in chapter 6 that damage to certain areas of the brain can give rise to a condition called pain asymbolia in which the subjective experience of pain is disrupted. Pure conscious intentions can be evoked by directly stimulating the brain. Patients who suffer from Alien Hand Syndrome have lost the conscious feeling of authorship generally associated with voluntary actions. In another impressive feat of "mind reading," researchers working on the neural basis of consciousness have recently developed a technique that allows them to quantify the degree to which people are conscious. And, as Dinesh D'Souza might be surprised to learn, this technique relies on physical activity in the brain.

More precisely, the technique relies on the brain's electrical response to a transcranial magnetic stimulation pulse (we discussed the use of TMS in the chapter 6). Based on this response, researchers can calculate a numerical index called the Perturbational Complexity Index (PCI), ranging from 0 (no complexity) to 1 (maximal complexity). This approach is particularly promising because it is noninvasive and can be used to reliably measure the degree to which people are conscious. In their work, the international team that pioneered the technique found that PCI values range from 0.44 to 0.67 in healthy patients who are fully awake, but that the values fall to 0.18 to 0.28 for people who are asleep (nonrapid eye movement sleep).[21]

To find out whether "switching off" consciousness chemically using anesthetics would have the same effects on the PCI, the team assessed data from patients who had received different amounts of three drugs commonly used as anesthetics. In these patients, PCI values ranged from 0.13 to 0.31—the low, unconscious levels seen in people who are asleep. Next, the team looked at patients with brain damage, who exhibited various levels of consciousness. Those who were in a vegetative state were clearly unconscious (PCI values ranging from 0.19 to 0.31). Patients who suffered from locked-in syndrome, a condition in which the body is almost completely paralyzed but cognitive faculties remain intact, were clearly conscious (PCI values ranging between 0.51 and 0.62). Patients in minimally conscious states showed intermediate PCI values (ranging between 0.32 and 0.49). So much for D'Souza's claim that consciousness has nothing to do with brain activity.

In his book on the reality of the soul and the afterlife, D'Souza cheerfully invokes the possibility of parallel universes and other exotic locales to substantiate his extraordinary conclusions. Maybe in one of those parallel universes consciousness, indeed, appears to be independent from matter, as D'Souza claims. Meanwhile, here on earth, the facts point to a rather different conclusion. I will leave the last word to astrophysicist Neil deGrasse Tyson, whom we met in chapter 2: "There is no shame in not knowing. The problem arises when irrational thought and attendant behavior fill the vacuum left by ignorance."[22] In the final two chapters, we will turn to the vacuum left by the demise of the soul, the existential anguish to which it gives rise, and the problem we face when decisions that affect all of us are made on the basis of irrational beliefs.

Chapter 8

THE SUM OF ALL FEARS

To defeat scientific materialism and its destructive moral, cultural and political legacies.... To replace materialistic explanations with the theistic understanding that nature and human beings are created by God.
——The Wedge Document, Discovery Institute, 1999

Many people find the idea of a soulless person quite unappealing, if not downright frightening. In the eighteenth century, La Mettrie's contemporaries regarded his radical materialism as such an insult to human dignity that the author of *L'Homme Machine* had to leave France to find refuge in first the Netherlands and then Prussia. At the dawn of the twenty-first century, this fear and indignation persist. In 1999, these sentiments were expressed in a controversial manifesto that surfaced on the Internet. According to the Wedge Document, issued by the Discovery Institute, the goal of a new generation of cultural warriors in America was to "defeat scientific materialism and its destructive moral, cultural and political legacies" and "to replace materialistic explanations with the theistic understanding that nature and human beings are created by God."[1] This message, incidentally, is consonant with what many in the United States believe today.

A good way to gauge the breadth of antiscientific sentiment in America is to pay attention to what our political leaders tell us. In 1999, for example, the year of the Wedge Document, America witnessed the horrifying shooting at Columbine High School in Colorado. Reflecting on this shocking event and the causes of youth violence, Tom Delay, then Republican House Majority Whip, pointed his finger at working parents who put their children into daycare, women who take birth-control pills, and the teaching of evolution in the schools.[2] According to Delay, we should not be surprised by such depraved acts of violence because the schools teach our children that they are "nothing but glorified apes who are evolutionized out of some primordial soup of mud."[3]

In 2012, the mass shooting at Sandy Hook Elementary School in Newtown, Connecticut, brought back the haunting memories from Columbine. As the country tried to wrap its mind around the senseless carnage, former Republican governor of Arkansas and 2008 presidential hopeful Mike Huckabee explained that we shouldn't be surprised by such events because we have systematically removed God from our schools.[4] These remarks by our political leaders are extreme, some would say utterly outrageous, but they nevertheless reflect a sentiment shared by many Americans: that the theistic view of nature and human beings is good, conducive to social harmony, and above all allows us to maintain our dignity. By contrast, the materialistic alternative offered by science is seen as sterile, dehumanizing, destructive, and something that should be opposed at all cost.

Closer to home, students have often asked me why anyone should bother living if people are indeed soulless. What would be the point? they ask. One of the questions in the soul-searching questionnaire from chapter 2 asked Rutgers undergraduates to describe what they couldn't do or be if they didn't have a soul. Students responded that they would lack feelings, compassion, the ability to love, the capacity to behave morally, and the freedom to make decisions. Even a professional scholar like Mark Baker, whose ideas we discussed in chapter 4, finds the idea of a purely materialistic world dizzying. Reflecting on what the demise of the soul would entail, Baker writes: "Ockham's razor is all very well, and it has its place as a useful tool, but I for one do not want to cut my throat with it and bleed to death."[5]

Are these fears legitimate? Is the scientific worldview as gloomy as its detractors portray it? Have we opened Pandora's box by exorcising the soul? Did we commit the mistake that Barrington Bayley tried to warn us about in *The God Gun*? By letting go of the soul, are we now doomed to live in Harry and Rodrick's world, a world devoid of meaning, purpose, and beauty? Perhaps even worse, would the widespread acceptance of the idea that people are soulless cause our society to plunge into chaos? These are the questions we face at the end of our soul-searching journey. In this chapter, I want to show you that the answer to these ominous questions is a resounding "no." Echoing the words of Charles Darwin, I will show you that there is indeed grandeur in the scientific view of life—and death—and that nothing gets lost morally, spiritually, or aesthetically by giving up our soul beliefs.

A good way to illustrate this empowering conclusion is through the story of Dumbo's magic feather.[6] In Walt Disney's famous story, Dumbo the elephant receives a magic feather from his friends that he thinks allows him to fly. In reality, the feather is just a psychological trick to boost Dumbo's confidence. He can fly perfectly well without the feather, relying on his natural abilities and his unusually large ears. In the final scene, Dumbo is poised to perform a perilous jump from an elevated platform. During his descent, he loses his magic feather and fears he will not be able to survive his last dive. In the nick of time, his friend Timothy the mouse manages to tell him that the feather is not magical and that he can fly on his own. Empowered by these words, Dumbo bravely deploys his oversized ears and flies around the circus in front of a dazzled crowd. What I want to show you in the pages ahead is that we are just like Dumbo—a bit superstitious, but brave and resilient nonetheless. We do not really need the soul, our magic feather, to soar above the abyss of nihilism and thrive as human beings. Let me play the role of Timothy the mouse, Dumbo's friend and ally, and show you why we can safely let go of our soul without fearing that the sky will fall on our heads.

THE STUBBORNNESS OF FACTS

If you suffer from soul-related anxiety, let me offer my first pill. To alleviate your worries, I'm going to ask you to do a little bit of thinking with me and consider the distinction between observations and their associated explanations. Take, for example, the behavior of physical objects like, say, apples. If you hold an apple in your hand and release it, it will fall to the ground (unless, of course, it is attached, supported by a table, and so forth). This is simply a *fact* about released apples, or similar physical objects. An important question, one that has kept history's geniuses at work, is how we can *explain* this fact. Here, we can see that the explanation has changed over time. Aristotle believed that released apples fall to the ground because they tend toward their natural place. In the seventeenth century, Isaac Newton dispensed with Aristotle's teleological (goal-oriented) account and argued that physical objects are subject to the goalless force of gravity, which he quantified mathematically. In the early twentieth century, the explanation changed again when Albert Einstein showed that the force described by Newton follows from the geometry of space-time.

What we see then is that while the *explanation* for why apples fall to the ground changes over time, the *fact* that apples fall to the ground does not change. Released apples fell to the ground in the time of Aristotle, Newton, and Einstein, and they will continue to do so in the future (at least as long as things don't get too ugly in the universe). Now imagine that someone were to express concern over the demise of earlier accounts of the behavior of apples. Suppose our friend Wayne, from chapter 3, worried that since Aristotle and Newton were wrong, apples, along with smartphones and all our other precious material possessions, will start flying off into the sky. This would of course be silly. The *explanation* for the behavior of apples may have changed, but *facts* about falling apples haven't. Apples do not care whether Aristotle, Newton, or Einstein is right; they will continue falling to the ground when released.

We can now transpose the analogy to human beings. An important *fact* about people is that we are sentient creatures capable of experiencing a broad range of psychological states. We fall in love or are moved by beautiful music; we can be jealous of our neighbors, saddened by a piece of news, or excited at the prospect of starting a new job. We can also make choices, behave morally, express our thoughts through language, create new mathematical theories, or write poetry. How can we account for the richness of our mental lives and the flexibility of our behavior? The traditional explanation used to be that this is all achieved by the immaterial soul. Recall that life itself was believed to emanate from the soul. But the *explanation* has changed. And so, just like people shouldn't worry that apples and smartphones will fly off into the sky because Aristotle or Newton were wrong, there are also no reasons to worry about the demise of the soul. Are people *really* going to stop falling in love or being moved by beautiful music just because mainstream scientists have reached the conclusion that there is no soul? Is everyone you know going to drop dead because life souls were eliminated from our scientific vocabulary? Of course not. The soul was only an *explanation* (or at least an attempt at one) for the *fact* that we are alive and have complex mental lives and flexible behavior. The explanation has changed, but the facts are here to stay.

Once the distinction between facts and explanations is brought into proper focus, it is easy to see that in losing the soul we lose only a potential explanation for some of the facts we hold dear about ourselves. I say potential explanation because, truth be told, the soul has never really explained much of anything, as we discovered in earlier pages. A more accurate assessment would be that the soul has been, and continues to be, the word we use to piously clothe our ignorance and give it an air of seriousness. For the ancients, everything was mysterious, and so we had life souls, free souls, body souls, ego souls, and surviving souls, as we saw in chapter 2. Today, we are repeatedly told that consciousness is the last frontier, as we saw in chapter 7, and so that qualia are the domain of the soul. But then again, saying that the soul gives us qualia is exactly the same as saying that we do not have a clue how qualia come to be. So if we remove the word *soul* from our vocabulary, and ditch the concept along the way, all we are left with is our ignorance. This way we can at least be more honest with each other. Instead of saying that the soul gives us this or that, we can simply admit that we have no idea how this or that actually works. Explanations wax and wane, but facts are stubborn and they are here to stay. By letting go of the soul, we haven't lost anything yet.

I said "yet" because things are never that simple. Albert Einstein wisely recommended that we should make things as simple as possible but not simpler. So let's follow the great man's lead. The demise of the soul, it turns out, poses a set of real problems to which we shall now turn. These problems, unlike the apparent one we just resolved, cannot be rectified with a simple analogy. Belief in the soul has given people the impression that we possess magical powers in addition to our more mundane capacities: a particular kind of free will and, of course, immortality. At worst, abandoning the soul would seem to entail abandoning free will and immortality. These are not trivial matters. For dualists like Dinesh D'Souza, the stakes could not be higher, and in a way I agree with them. Where I disagree with D'Souza and the other New Dualists is on how we should resolve the issues.

In the pages ahead, I will show you that the kind of soul-based free will that dualists insist we must have is another chimera. Not only is that sense of free will incoherent, but we should consider ourselves damn lucky that free will isn't actually soul-based. As for immortality, the stakes are in the eye of the beholder. Not being immortal may suck for many people, but it really doesn't have to be that way. Soul beliefs represent one way of dealing with our own mortality, but it is certainly not *the only way*. As I will show you, people who do not believe in the soul narrative can live happy lives without losing sleep over the prospect of their own demise. In the end, even if we take into account the problem of free will and death, we can maintain our conclusion that there is nothing to lose by letting go of the soul.

SOUL-FREE

The question of free will is one of the oldest and toughest philosophical chestnuts. It has been pondered, analyzed, and dissected by history's deepest thinkers, but no one has ever been able to resolve it to everyone's satisfaction. The reason free will is so important is that it lies at the heart of everything that matters to us—from personal feelings of pride and regret to broader issues of moral responsibility, criminal law, politics, and religion. Before giving you a feel for the terrain, an important disclaimer is in order. Free will has been discussed for more than two thousand years, and I would be a fool to pretend that I can settle the question, propose a novel account, or even say anything new about the topic. I simply cannot do that, and if anybody else can, then I'll be the first one to applaud. My goal is much more modest. I simply want to show you how the question of free will is related to soul beliefs and consider what happens to our understanding of free will if we remove the soul from the human equation. The answer, my second pill against soul-related anxiety, will be that any account of free will based on the soul is a nonstarter and that whatever free will is we should consider ourselves lucky that it is not soul-based.

So what exactly is the problem of free will and how does it relate to the soul? The best illustration I have seen is in a Dilbert comic strip by Scott Adams.

Adams's cartoon depicts a conversation between Dilbert and his anthropomorphized dog, Dogbert. Dogbert asks whether Dilbert believes brain chemistry controls people's actions, and Dilbert says yes. Dogbert then asks him how we can be held responsible for anything if brain chemistry controls what we do. Dilbert explains that people have free will and can choose what to do. Inquisitive Dogbert then asks whether free will is separate from the rest of the brain and whether it is free from the laws of physics. Dilbert tells him it is not separate, but rather the part of the brain that is free, but that, yes, it must be free from the laws of physics or we wouldn't be able to hold people accountable for their actions. In the last panel, Dogbert asks how the "free will" part of the brain is connected to the rest—the part that *is* affected by the laws of physics. Dilbert, knowing he can't answer that question to anyone's satisfaction, simply tells Dogbert to shut up.

Dilbert's intuition is that in order to be free, our actions cannot be constrained by the laws of nature. This conception of free will is called libertarian (nothing to do with the political philosophy) or contra-causal free will. According to *Theopedia*, *libertarian free will* means that "our choices are free from the determination or constraints of human nature and free from any predetermination by God."[7] Moreover, "All 'free will theists' hold that *libertarian freedom* is essential for moral responsibility, for if our choice is determined or caused by anything, including our own desires, they reason, it cannot properly be called a free choice. Libertarian freedom is, therefore, the freedom to act contrary to one's nature, predisposition and greatest desires. Responsibility, in this view, always means that one could have done otherwise."

If you were to ask a believer in libertarian free will what causes her (voluntary) actions, she would answer that her will does. And if you asked her what causes her will, the answer would be: nothing. The buck stops there. Our will causes our actions, but it is not itself caused by anything—it is free. Advocates of contra-causal free will believe that human beings are prime movers, or uncaused causes. Now, what could possibly give us such powers? What would be able to transcend the laws of cause and effect described by science? The immaterial soul, of course! If our actions were the product of the unfathomably complex sum of all the causal factors impinging upon us, including the influence of our genes, our environment, the chemistry of our brains, and the laws of physics, then we would be fully caused creatures lacking metaphysical freedom (at least to a believer in libertarian free will). But thanks to our soul, which can escape the influence of any causal factor known to science, our will is free. Because the soul can causally interact with the body, it can make us do what we do, while itself remaining uncaused, and therefore free (see figure 8.1).

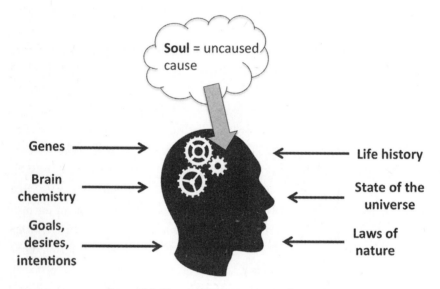

Figure 8.1: The soul as an uncaused cause.

Among believers in contra-causal free will, we find philosophers,[8] some of the New Dualists we met in previous chapters, and, according to other philosophers, the majority of regular folks and even children.[9] In the soul survey I described in chapter 2, we asked our participants what they thought it meant to have free will by giving them the choice between two definitions, one contra-causal and the other not. Here's the non-contra-causal definition:

- Having free will means that even though my choices—and thus my will—are ultimately caused by factors such as my genes, my upbringing, what goes on inside my brain, and what goes on in the outside world, I am free to act on those choices if I wish to do so.

And here's the contra-causal one:

- Having free will means that my choices are ultimately free from the determination of any physical cause, including my genes, my upbringing, what goes on inside my brain, and what goes on in the outside world. In other words, while my decisions are caused by my will, my will itself is not caused by anything—it is free.

We found that a quarter of our participants, 26 percent, selected the contra-causal definition. We also found that the value on the soul scale (ranging from 1 through 6) selected by our participants was positively correlated with the value they selected on the free-will scale (also ranging from 1 through 6).

If we let go of the soul, as we should, we also give up the mechanism for contra-causal free will. Is this something we should worry about? There are two issues we need to consider here. The first is that according to soul-based free will, our will is free in the sense that it is not caused or constrained by anything at all. The second issue is that the contra-causal notion of freedom is essential for moral responsibility. If we abandon the soul, then must we give up the notion of free will altogether and do we also stand to lose any meaningful notion of moral responsibility?

To make things more tangible, let us consider a concrete example and play the game that believers in contra-causal free will would like us to play. This will bring the key issues into sharper focus and will allow us to better assess the situation. Here's a hypothetical example I constructed for this purpose. Imagine a gigantic boulder sitting on top of a hill. At the bottom of the hill, there is a small town composed of few hundred homes. The boulder has been on top of the hill for centuries, but over time erosion, the weather, and the effect of rising temperatures have made it unstable. One night, during a powerful storm, the combined effects of the rain and the wind give the boulder a final push and cause it to come crashing down the hill, flattening many of the homes in its destructive path and killing dozens of people. Let us now ask the critical question for believers in libertarian free will: Could the boulder have done otherwise? The answer in this case is clearly no. The boulder was subject to forces such as erosion, the wind, the rain, and gravity, which it had no control over, and so it did the only thing it could do under those circumstances—come crashing down the hill.

Imagine now a troubled youth who desperately needs money and decides to rob a convenience store. Masked and armed, the young man walks into the store and demands the content of the cash register. The clerk resists, the two men argue, the tension escalates, the young man panics, and he shoots and kills the clerk. We can now ask our critical question again: Could the young man have done otherwise?

Many people, including some of the students I have asked, will tell you that the young man could indeed have chosen to do otherwise and not pull the trigger. Intuitively, this feels like the right answer, and it is precisely this intuition that believers in contra-causal free will rely on. They want to convince us that their view of freedom is intuitively correct and that if science clashes with it, then there must be something wrong with science. Now why would science clash with the intuitive view that the troubled youth could have done otherwise? To understand why, let us think about the young man's predicament a little more carefully. A more careful analysis of the situation tells us that the store clerk was shot dead because the young man pulled the trigger. This is of course a truism, but it is also what happened.

From this starting point, we can reason backward and ask what caused each of the events that led to the clerk's death. If the trigger was pulled, it was because the young man pressed it. The pulling of the trigger occurred because the young man's shooting finger received appropriate nerve signals from his motor cortex. Those signals were themselves the product of unconscious neurochemical activity in the young man's brain. Consciousness may have intervened in the process but it too, as we saw in previous chapters, is the result of unconscious neural activity. The unconscious neural activity in the young man's brain, in turn, was determined by the complex interplay of genetic influences, environmental factors (including the young man's life history), and the operation of physical laws.

We could be very diligent, ask about the causal antecedents of the three sets of influences I just described, and follow the causal chain all the way back to the moment the universe began, but that won't be necessary. It should now be plain to see that at the moment of the shooting, the young man had no more control over the genes he inherited from his parents, his life history, or the laws of nature than our hypothetical boulder had control over the rain, erosion, or gravity. If the causal chain that led the young man to pull the trigger runs unbroken and is ultimately anchored in aspects of the world over which he has no control, then how could he possibly have done otherwise?

This conclusion may be counterintuitive, but it inescapably follows from the fact that we are purely material creatures living in a material world. Like boulders, rabbits, black holes, and galaxies, we are an integral part of the causal nexus. Without a contra-causal soul to rely on, we simply cannot escape the laws of cause and effect. And that's precisely the problem for someone who believes in contra-causal free will. Our intuitions tell us that the young man could have done otherwise, and so he must be morally responsible for his actions. On the other hand, a more reasoned analysis reveals that the young man could not have done otherwise, and therefore, we seem forced to conclude, he cannot be held morally accountable—or so it would seem.

What I just described is the classic problem for free will posed by determinism. If the world is completely determined by past states of the universe and the laws of physics, then libertarian free will is an illusion. One way out of this impasse would be to argue that determinism isn't true. Perhaps quantum mechanics, the branch of physics that deals with the behavior of atoms, injects true randomness into the world. Maybe so, but this would not solve the problem. Even if we allowed for the effect of subatomic coin flips intervening at some point in the causal chain, this does not change the situation. The young man is no more in control of random processes than he is in control of factors like his genetic makeup, his upbringing, or the unconscious neurochemical activity in his brain. Besides, nobody would want to say that the young man's choice to pull the trigger was "free" because it was random.

We seem to really have reached an impasse then. Both the boulder and the young man are causally responsible for the death of their respective victims, and neither the boulder nor the young man could have done other-wise. This is where believers in contra-causal free will begin to panic. If we do not hold the boulder morally responsible for the death of its victims—after all, we do not put boulders on trial for murder—then why should we hold the young man morally responsible? they would ask. Both the boulder and the young man could not have done otherwise, their behavior was fully caused, so how can we exempt one from moral responsibility but not the other? The problem we now face, believers in contra-causal free will would argue, is that if we remove the soul from the human equation, and thus dispense with libertarian free will, we can no longer hold people accountable for their actions and we lose any meaningful notion of moral responsibility.

At first blush, the problem seems valid. Upon closer examination, however, we will see that it is a false problem. To understand why, let us begin by asking ourselves whether our choices really need to be uncaused in order to be free—the key assumption underlying libertarianism. Reflecting on this approach to the problem of free will, David Hume, the eighteenth-century Scottish philosopher, already smelled a rat. Hume concluded that unless our actions are indeed *caused* by our desires, temperament, dispositions, habits, inclinations, personality, and so forth, they would not make any sense. In Hume's own words, "Actions are, by their very nature, temporary and perishing; and where they proceed not from some cause in the character and disposition of the person who performed them, they can neither redound to his honour, if good; nor infamy, if evil."[10]

Imagine, for example, that I invited you to my house, dear reader, so that we could discuss the issue of free will over a drink. Suppose now that in the middle of our conversation, and completely out of the blue, I threw my martini in your face. Would you not immediately ask yourself why I did such a thing? Was it something you said to me? Do I suffer from Alien Hand Syndrome? Do academics like to shower their guests with vodka and olives? Surely you would look for a *cause* or a *reason* to try to explain my baffling behavior. If after you asked me I told you that my decision to throw my martini in your face was caused neither by my goals, my desires, my state of mind, nor by anything else—if I told you that it was a pure manifestation of my metaphysically *free* will—you'd probably conclude that I am insane, not free.

Fortunately, we do not live in a world in which people's actions are completely unconstrained and utterly capricious. Every time I get behind the wheel, I am relying for my own survival on the fact that the will of other drivers is constrained by their desire to not get into an accident, their goal to reach their destination safely, and their intention to continue living for at least another day. If the world were populated by contra-causal souls, you would be taking astronomical risks every time you got behind the wheel, not so much because of a few reckless drivers, but because of all the other perfectly normal drivers who could at any instant "freely" decide to pulverize your car just because their will is "free" and therefore not constrained by the rules of elementary sanity. Worse even, if people really had souls and contra-causal free will, we could kiss our cherished values and the institutions that depend upon them good-bye. What would be the point of teaching children right from wrong, of trying to deter or rehabilitate criminals, or even trying to reason with anyone, if people's souls could do as they damn well pleased anyway?

Hume's conclusion is that our choices do not need to be uncaused in order to be free—quite the contrary. In his book *Elbow Room*, Daniel Dennett concludes that contra-causal free will is not a kind of free will worth wanting. He also shows that the key question for libertarians, the question of whether someone "could have done otherwise" *in exactly the same circumstances*, is meaningless because, for all intents and purposes, it is a question to which there is no answer knowable to us. Suppose you decided to buy this book at some point in the past. If you believe in contra-causal free will, you must also believe that if you were placed back *in exactly the same situation*, you could have decided to do otherwise and not buy the book. But how could you possibly know that? Sure, you may have a hunch or an intuition about it, but if an issue as monumental as moral responsibility hinges on the answer to that question, we would want to rely on something a little more solid than a hunch or an intuition.

Now if you went back to the bookstore tomorrow and placed yourself in the same situation, you could decide not to buy the book again and then claim that you can indeed do otherwise. But this would be completely irrelevant. Of course you can do otherwise when the situation is similar but not exactly the same! The second time you went to the store, you were not *in exactly the same situation* that you were in the first time you went to the store. For one thing, when you returned to the store, you had read this passage from the book (something you were unlikely to have done the first time around, since you hadn't bought the book yet!), and that's what could have motivated you to not buy the book a second time. If we compare apples to apple-looking oranges, then yes, anybody can do otherwise. But if we compare apples to apples, then nobody can really tell whether they would have been able to do otherwise—and that's the problem.

In sum, contra-causal free will is muddled and so we do not need or want it. How about the problem of moral responsibility? If science tells us that our actions are caused by forces beyond our control, doesn't that entail that we should all be blameless, including the worst criminals, just like the boulder in my earlier example? Fortunately, a moment's reflection suffices to show that it would be utter madness for anyone to suggest that we should free all the rapists and murders simply because they could not have done otherwise and are therefore blameless victims of nature. The real problem lies not with what science is telling us about the causes of human behavior but rather with the logic underlying contra-causal free will. Libertarian theists insist that human beings have magical powers—contra-causal free will—and that moral responsibility critically depends on those powers. (Don't forget the story of Dumbo and his feather!) No magical powers, they argue, no moral responsibility. But this line of reasoning, in addition to relying on magic, makes the untenable assumption that there is no nonmagical capacity possessed by human beings—but not by boulders—that could serve as an anchor for moral responsibility.

Here's a nonmagical capacity possessed by human beings but not by boulders: a broad sense of rationality. By this I mean something benign and uncontroversial. I am merely referring to the fact that healthy adults have the capacity to formulate plans and goals; to share them with others through the vehicle of language; to consciously reason about the consequences of their actions; to understand the difference between right and wrong; to be sensitive to the suffering of others; to adjust their behavior based on new information or factors such as praise, credit, blame, or punishment; and to treat other human beings as autonomous moral agents endowed with similarly rational minds.

With this view, our minds may be fully caused, but this is precisely why we can be rational in the first place (again, in the benign sense just described), so determinism is a good thing after all. We have the rational powers I just described *because* our minds are rule-governed. Unlike libertarianism, which views free will and determinism as incompatible, I am arguing for a compatibilist view of moral responsibility—one where determinism and responsibility are indeed compatible. Needless to say, this road has been traveled before. David Hume arguably provided the most influential statement of compatibilism and, more recently, Daniel Dennett also championed this position.

Let me now show you that this approach to moral responsibility offers a perfectly respectable alternative to soul-based responsibility—in fact, a much better one. First, the rationality-based view offers a naturalistic account of moral responsibility that does not face any of the nonsensical implications that beset contra-causal accounts (me throwing my martini in your face for no reason, someone "freely" crashing into your car, and so on). Second, by tying moral responsibility to capacities that people really possess, and not to alleged magical powers, our account securely attaches responsibility to *facts* about human beings that will be immune to changes in possible explanations. Third, our account is in accord with Hume's important conclusion that it would be unjust to hold someone morally accountable for his actions unless these actions were caused by his desires, goals, intentions, plans, and so on.

Fourth, the rationality-based account corresponds precisely with what the law cares about for purposes of ascribing responsibility. In the United States, a person is found guilty of a crime if she performed the relevant action, if the action violates a criminal statute, and if the action was performed with a criminal state of mind—the famous *mens rea* clause ("guilty mind"). The mind is guilty if the action was performed knowingly, willfully, and if it involves gross negligence or reckless indifference. This ties in with Hume's insistence that we distinguish between voluntary and involuntary actions (the former being the relevant ones for purposes of moral responsibility). There is indeed an important difference between someone who willfully steps on your foot and someone who merely does so unintentionally.

Fifth, the rationality-based approach also provides an adequate basis for diminished responsibility. The reason we do not judge children and the mentally handicapped the same way we do healthy adults is that rationality is a natural capacity that is not yet fully developed in children and that can be compromised in the mentally ill.

Finally, the rationality-based approach is robust enough to handle metaphysical worries that a deterministic world would mean our choices were not really our own. Dinesh D'Souza expresses this worry when he writes, "In the absence of free will, none of the decisions that you believe you made in your life were actually made by you."[11] This is simply not true! If there is no soul, there is no contra-causal free will, and there is no immaterial "you," but there remains a physical "you." And it is this "you" who makes the decisions. If decisions are made by your brain instead of your soul, I do not see how someone could claim that they are somehow not made by "you." Our brains are powerful rational engines capable of making decisions precisely because they are governed by deterministic principles. We have the ability to consciously predict the likely consequences of our actions. Armed with these predictions, we can select the options that appear the most likely to allow us to reach our goals.

This is rationality in action. As intelligent beings, this is how we make our decisions. Instead of being uncaused causes, which as we saw, would have nonsensical implications, we are caused causes. So what? Regardless of whether the future is in principle predictable from past states of the universe and the laws of physics, it is certainly not predictable by us. The future therefore remains epistemically open to us. But Mother Nature endowed us with the ability to predict the consequences of our actions, actions that do have real consequences in the world. The more we understand, the better our limited predictions become, the sweeter the future will be for all of us. Now's certainly not the time to sit back and complain that there is no point in doing anything because everything is already determined. Fatalism is for losers.

We can now return to the two problems that motivated this discussion. If we give up the soul, do we lose our conception of free will and can we maintain a meaningful notion of moral responsibility? The answer I gave is that we only lose a *kind* of free will, the contra-causal type, a variety of free will that is not even worth wanting in the first place. For someone like Daniel Dennett, the demise of libertarian free will does not entail the loss of other kinds of free will. For someone like Sam Harris, no contra-causal free will means no free will, period. Whatever the correct answer may be, to the extent that there even *is* a correct answer to such a question, the demise of the soul and the concomitant loss of libertarian free will do not erase moral responsibility. As we saw, there is a perfectly respectable notion of moral responsibility anchored in a capacity that human beings really possess—a broad sense of rationality. This conception of moral responsibility seems to be able to deliver everything we need to continue functioning without fearing that the sky will fall on our heads. In the domain of freedom and responsibility, we lose nothing by giving up the soul.

TILL DEATH DO US PART

In his foreword to Dinesh D'Souza's book *Life after Death: The Evidence*, evangelical pastor and megachurch founder Rick Warren offers a sobering assessment of what life would be like without the soul and the promise of immortality. "If this life is all there is," he writes, "there is no basis for any meaning, hope, purpose, or significance to life. . . . The logical end of such a life is despair."[12] Warren's comments, whether realistic or not, stem from a perplexing aspect of the human condition that philosopher Stephen Cave calls the Mortality Paradox. On the one hand, due to our sophisticated cognitive capacity, we are a species, perhaps the only species, whose members are aware of their own mortality. On the other hand, it is very hard for us to imagine what it would be like to be dead. If you are picturing yourself in a completely dark and lonely place, you are not trying hard enough. You are still there to see yourself in that predicament. The truth is that consciousness simply cannot simulate the state of being dead any more than it can help us visualize ourselves before we began existing.

We know that we must die, and yet, at the same time, being dead is unimaginable to us. What the soul narrative provides, according to Cave, is a way to resolve the Mortality Paradox. Soul believers accept the first part of the paradox as they too realize that our bodies must one day perish. By claiming that the real person is the immortal soul, believers resolve the second part of the Mortality Paradox. If the soul carries consciousness into the afterlife, then we do not have to worry about ever ceasing to exist. But if we abandon the soul, opponents of scientific materialism warn us, then the Mortality Paradox will become paralyzing. For Rick Warren, a soulless existence would lead to despair.

What I would like to show you in the pages ahead is that there is absolutely no reason to react with doom and gloom to the demise of the soul. We could take two different routes to reach this empowering conclusion. The first one consists of showing that the soul narrative doesn't in fact resolve the Mortality Paradox and that it may even exacerbate it. We will briefly consider this approach but conclude that it is not decisive. The other approach is to show that the soul narrative, whatever its merits may be, is not necessary to resolve the Mortality Paradox. If there are indeed other ways to resolve the Mortality Paradox and alleviate the fear of death, then the soul narrative becomes superfluous, and we have no reason to mourn its demise. This is the path that we will follow.

To begin, let us examine the reasons that have led some authors to reject the soul narrative as a viable solution to the Mortality Paradox. In his book *Immortality*, Steven Cave concludes that possessing a soul would be a terrible curse. Cave's reasoning focuses on the psychological effects of eternity, what he calls the long dark tea-time of the soul (a reference to the Douglas Adams novel of that name). Cave argues that soul-based immortality faces two main problems. The first is the boredom and apathy that would come from having lived a very long time, from having done everything there is to do and experienced everything there is to experience. The second is the paralysis associated with the realization that the time that we still have ahead of us after having lived such a long life is infinite. Cave concludes that these two problems—the backward-looking and the forward-looking—"threaten to suck the meaning out of life and leave one wishing for a terminal deadline."[13]

This is exactly the opposite of Rick Warren's conclusion. If Cave is right, soul-based immortality would be a terrible curse indeed. To illustrate the tragedy of immortality, Cave recounts a story by the Argentinean writer

Jorge Luis Borges called *The Immortal*. The story is about a Roman soldier and his quest to find the fountain of youth. Near the end of his journey, the soldier comes in contact with cave-dwelling creatures, the Troglodytes, who live in complete indifference to their surroundings. The supreme irony of Borges's story is that the withered and apathetic Troglodytes are the immortals. They began their voyage through eternity animated by the desire to build the city of their dreams. But as boredom and pointlessness set in, their architectural project turned into an absurd construction that they later abandoned to become Troglodytes in the desert.

In *The Portable Atheist*, the late Christopher Hitchens takes a similar approach and argues that the vision of eternity offered by the soul narrative would in fact be a nightmare. In Hitchens's own, colorful prose: "Who wishes that there was a permanent, unalterable celestial despotism that subjected us to continual surveillance and could convict us of thought-crime, and who regarded us as its private property even after we died? How happy we ought to be, at the reflection that there exists not a shred of respectable evidence to support such a horrible hypothesis."

Cave and Hitchens may very well be right in thinking that soul-based immortality would be a terrible curse. But it is difficult to know for sure. Besides, the soul myth can easily be adjusted and inoculated against the criticism that Cave and Hitchens level against it. As Cave himself acknowledges, the trick can be accomplished by switching from an anthropocentric view of the afterlife to a theocentric one. In the anthropocentric narrative, the afterlife is envisioned as an idealized version of life on earth. On this scenario, the departed can get together with their friends and relatives, be married, watch movies, go to shopping malls, have hobbies, and enjoy music and the arts. This is the vision of heaven offered today by some American evangelist pastors.[14]

The human-centered view of the afterlife could certainly lead to the kind of problems Borges tried to warn us about in *The Immortal*. But this can be easily fixed by switching to the theocentric view. This is the route taken by none other than Jesus himself, when he was asked an inconvenient question by a group of Jews who did not believe in the afterlife. The men wanted to know what would happen to a woman who had been married and widowed several times here on Earth. Who among her multiple spouses would end up being her husband in the afterlife? Jesus conveniently answered that the question was moot because people are like angels in the afterlife, and the notion of marriage does not apply to angels.[15]

Trying to deny that the soul narrative resolves the Mortality Paradox, the first line of defense to counter Rick Warren's nihilistic exhortations, does not strike me as having a very high probability of success, because this approach forces us into absurd discussions about life as a human being or an angel, the pros and cons of eternity, and other questions that rational discourse simply cannot settle. Fortunately, there is a more promising approach, one firmly anchored in reason and evidence. In asserting that the rejection of the immortal soul would rob our lives of meaning and lead us all to despair, Rick Warren makes a claim, however offensive it may sound to nonbelievers, that at least has the virtue of being testable.

To assess the merit of Warren's claim, all we need to do is shift our focus away from America and consider societies in which religious myths do not hold such a firm grip on people's minds. This is what Phil Zuckerman, a professor of sociology at Pitzer College, did in his fascinating book *Society without God*. Zuckerman spent a little over a year living in Aarhus, Denmark's second largest city. Zuckerman chose Denmark and neighboring Sweden because these Scandinavian countries are two of the least religious countries in the world. Here in the United States, as we saw in chapter 6, close to 90 percent of the population believes in God. In Denmark and Sweden only a minority of the population believes in God—24 percent of Danes and only 16 percent of Swedes, according to one study.[16] Belief in life after death, which ranges between 70 percent and 80 percent among Americans, is as low as 30 percent among Danes and Swedes. Zuckerman also considers belief in heaven and hell, religious service attendance, how often people pray, and the importance of faith in one's life, and he shows that all these indicators of religiosity are at floor levels in Denmark and Sweden.

If religiously inclined conservatives like Rick Warren—but also Pat Robertson, Bill O'Reilly, Rush Limbaugh, and Ann Coulter[17]—are right, then Denmark and Sweden, societies where most people do not believe in God and the soul, should long have descended into chaos, and most Danes and Swedes should be utterly miserable. One of the central conclusions of

Zuckerman's research is that this is simply not true. Quite the contrary; by almost any standard imaginable, Denmark and Sweden are two of the healthiest, happiest, and most prosperous countries in the world. Every year, the United Nations ranks the world's nations on a Human Development Index, which measures average societal achievements on the basis of three main dimensions of human flourishing: longevity and health, access to education, and standards of living. Denmark and Sweden invariably fall in the group of countries with very high levels of human development. In 2013, the latest report available as of the writing of this book, Sweden ranked 7 and Denmark 15 on the Human Development Index.

In his book, Zuckerman compiles an impressive list of indicators, including measures of life expectancy, wealth, economic and gender equality, social justice, economic competitiveness, healthcare, environmental protection, lack of corruption, propensity to be charitable toward poor nations, murder rates, and individual happiness, which all point to the same conclusion: Denmark and Sweden are two of the safest, healthiest, fairest, and happiest nations in the world. Zukerman concludes that if there is an earthly heaven for secular folk, Denmark and Sweden would be it: "healthy democracies, among the lowest violent crime rates in the world, the lowest levels of corruption in the world, excellent educational systems, innovative architecture, strong economies, well-supported arts, successful entrepreneurship, clean hospitals, delicious beer, free health care, maverick filmmaking, egalitarian social policies . . . and not much faith in God."[18]

The existence of countries like Denmark and Sweden torpedoes claims by religious apologists like Rick Warren that without belief in God and the immortal soul human beings would be reduced to a life of despair. Belief in God, the soul, and the afterlife are simply not necessary for human beings to flourish. Zuckerman devoted a chapter of his book to death and the meaning of life to find out how the Danes and the Swedes deal with the Mortality Paradox. Here's what Anders, a forty-three-year-old father of two from Aarhus replied when Zuckerman asked him what he thought would happen after he died: "I think we just die. That's why we have to have our lives when we've got it . . . before I know—it's over. So you just have to live every day . . . and make nice days. And I try to do that, actually. I have a good time here and I've got a good family. I've got some good friends, and . . . I'm a very lucky person."[19]

Much to Zuckerman's surprise, attitudes like Anders's were very common among the many Swedes and Danes he interviewed. Few of them seemed to lose much sleep over the Mortality Paradox. When Zuckerman asked Leif, a seventy-five-year-old publisher, what he thought happened after we died, Leif replied, "Nothing."[20] Zuckerman asked him how that made him feel, and Leif replied, "Well, not very sorry. It is as it is. Really, I don't feel anything about it especially." Zuckerman concludes that many people in these countries are able to live their lives perfectly normally "without great fear of, or worry about, the Grim Reaper."[21] One of the people Zuckerman interviewed was a forty-three-year-old hospice nurse from Aarhus. In her many years of experience with people who were about to die, the nurse found that it was actually atheists who had the easiest time accepting their fate, while Christians had a much more difficult time facing death.

Swedes and Danes have found a way to resolve the Mortality Paradox without recourse to the soul narrative and while being able to maintain perfectly happy lives. Work by Ronald Ingelhart and his collaborators shows that the Swedes and the Danes are indeed some of the happiest people in the world. These researchers asked people from forty different countries the following question: "During the past few weeks, did you ever feel on top of the world/feeling that life was wonderful?" The country with the highest percentage of people who responded "yes" to that question was Sweden, with Denmark not far behind in third place.[22] More generally, Zuckerman's research clearly demonstrates, without having to speculate about the fate of angels or the psychological effects of eternity, that the soul narrative is not necessary for human beings to flourish and live happy, meaningful lives. In the next chapter, I will show you that scientific materialism, so feared here in the United States, actually provides a solid basis upon which to build an alternative, naturalistic narrative that allows us to find meaning in life and effectively cope with the Mortality Paradox.

In life as in death we have nothing to lose by letting go of our soul beliefs. When I say *we*, whom exactly am I referring to? As I've said before, I am not writing this book to bully people who believe in the soul into letting go of their cherished doctrine. If a mother who has just lost a child finds comfort in the idea that his soul is in heaven and that the two of them will one day be reunited, I wouldn't have the slightest inclination or the arrogance to try to tell her otherwise. What we have to worry about in denying the existence of the soul is not what would happen to people who do believe in its existence. Most of these people will probably continue to believe no matter what. They are the Agent Mulders I was referring to earlier. And if believing in the soul brings them comfort and peace, then so much the better.

The real question to ask is what would happen if no new person born in the United States as of tomorrow were to believe in the soul. Existing soul-believers would be grandfathered in, so to speak, but no new soul believers would be born. Within a few generations, and without bullying anyone, we would end up with an entire country of people who do not believe in the soul. Would all these soul-free people necessarily be doomed to a meaning-less life of despair, as people like Rick Warren would have us believe? The example of the Swedes and the Danes, and of all the nonbelievers here in America, demonstrates that human beings do not necessarily need the soul narrative to deal with the Mortality Paradox. Believing in the immortality of the soul may be one way to cope with death, but it is not the *only* way, or even the most efficient way. If so, "we," a hypothetical future nation of non-believers in the soul, would have nothing to lose.

THE QUESTIONABLE VALUE OF BELIEF

If you listen to the New Dualists, they will tell you that their belief system is not only supported by credible scientific evidence but that it is also bene-ficial to individuals and society. This is what Dinesh D'Souza asserts in his book on the afterlife. He assures his readers that "there is strong evidence that belief in life after death makes your life better and also makes you a better person."[23] D'Souza also maintains that belief in life after death is a conviction "that sustains and strengthens our civilization,"[24] and he trium-phantly concludes that "unbelief is neither intellectually plausible nor prac-tically beneficial."[25] If these claims were really true, then we would have something to lose by giving up our soul beliefs. However, Phil Zuckerman's work on countries like Denmark and Sweden should make us suspicious of D'Souza's sweeping generalizations. And if we take into account the full set of data available on these questions, the positive effect of religious beliefs touted by New Dualists like Dinesh D'Souza either evaporates or reduces to factors that have little, if anything, to do with people's religious beliefs.

An easy way to test D'Souza's broad claims is to consider the cross-national pattern of relationships between levels of religiosity and levels of societal health in developed nations. This is what Gregory Paul did in a recent study published in the *Journal of Religion and Society*. Paul explains that the twentieth century acted as "a vast Darwinian global societal experiment" in which different models of social, religious, political and economic organization competed with each other.[26] This natural experiment gives us precisely the cross-national data we need to test D'Souza's claims about the positive relationship between religious belief and societal health. After reviewing the data, Paul concludes that "the highly secular democracies consistently enjoy low rates of societal dysfunction, while proreligious and antievolution America performs poorly."[27] He explains that, in general, "higher rates of belief in and worship of a creator correlate with higher rates of homicide, juvenile and early adult mortality, STD infection rates, teen pregnancy, and abortion in the prosperous democracies."[28] To take just one revealing example, rates of gonorrhea infection in adolescents are six times to three hundred times higher in the proreligious United States than in the other, more secular prosperous democracies.

In his article, Paul is careful not to conclude that elevated levels of religiosity are the actual cause of higher levels of societal dysfunction. Correlation does not entail causation. All the same, if the data showed that the proreligious United States enjoyed higher rates of societal health compared to the other more secular first-world nations, then D'Souza's claim about the practicality of belief would be supported. But this is not what the data show. As Paul remarks, "only the more secular, pro-evolution democracies have, for the first time in history, come closest to achieving practical 'cultures of life' that feature low rates of lethal crime, juvenile-adult mortality, sex related dysfunction, and even abortion."[29] In sum, the claim that strong theistic beliefs lead to healthier societies is not supported by the data.

In another recent article on a related set of questions, psychologist Paul Bloom reviews the literature on the effects of religion on people's prosocial behavior. For New Dualists like Dinesh D'Souza, it is *belief* in the immortality of the soul that is supposed to make people better and happier. After reviewing the published literature on this question, Bloom finds that there is little evidence supporting D'Souza's conclusion. As Bloom writes, "I conclude that religion has powerfully good moral effects and powerfully bad moral effects, but these are due to aspects of religion that are shared by other human practices. There is surprisingly little evidence for a moral effect of specifically religious beliefs."[30] To support this conclusion, Bloom makes a useful distinction between different senses of *religion*. When people talk about their religious experience, they may be describing a type of transcendental experience often characterized by a deep feeling of awe and a sense of oneness with the rest of the universe. Another sense of religion involves belief in the existence of supernatural entities like God and the immortal soul. Yet another sense of religion corresponds to a certain mode of social interaction, like attending church or taking part in other religious ceremonies.

The three senses of religion described by Bloom often coexist in people who call themselves religious, but they can be found independently of each other as well. For example, Albert Einstein often talked about the feeling of awe he experienced while engaged in his scientific work, but he also made it very clear that he did not believe in a personal God or the afterlife. In his book, Phil Zuckerman explains that the Swedes and the Danes are religious in the sense that most of them support their national churches through taxes, get married in church, and still choose to have their children baptized. But, as we saw, the Swedes and the Danes also tend not to believe in God or the soul narrative. In this case, the social aspect of religion continues to exist without the presence of supernatural beliefs. This is why Zuckerman called his book *Society without God* and not *Society without Religion*. This more nuanced understanding of religion advocated by Bloom allows us to expose the fallacy in D'Souza's reasoning. Just because someone demonstrates that "religion" has an effect on people's prosocial behavior, we cannot automatically conclude that the effect is driven by people's supernatural beliefs. Nevertheless, this is precisely the conclusion that D'Souza advertises in his book.

Consider, for example, laboratory studies in which people are primed by sentences containing religious words like *spirit*, *divine*, *God*, and *sacred*. In subsequent tests, the same people are found to be more generous in economic games in which they have to decide how much money they wish to give to other players. Based on such findings, it would be easy to conclude that supernatural beliefs are good for us because they make people more generous. Until you learn that the same effects can be obtained by replacing the religious words by secular alternatives like *police*, *jury*, *court*, *civic*, and *contract*.[31] In their book *American Grace: How Religion Divides and Unites Us*, based on two of the most comprehensive surveys on religious attitudes in America, Robert Putnam and David Campbell conclude that people's religious beliefs have little, if anything, to do with prosocial behavior. What really seems to matter is belonging to a community. As Putman and Campbell explain, "In fact, the statistics suggest that even an atheist who happened to become involved in the social life of the congregation (perhaps through a spouse) is much more likely to volunteer in a soup kitchen than the most fervent believer who prays alone. It is religious belongingness that matters for neighborliness, not religious believing."[32] Belonging to a social group and taking part in communal activities is something that anybody can do with or without religion. Once again, we have nothing to lose by letting go of our soul beliefs.

Chapter 9

IMAGINE

Imagine there's no Heaven ... and at once the sky's the limit.
—**Salman Rushdie,** *Letters to the Six Billionth*
World Citizen, **1999**

At the stroke of midnight, Clinias walked his friend Marcus to the temple's inner sanctum. "Soon, we will know what's on the other side," Marcus said in a calm voice. "Or I may never see you again," Clinias replied. Both men were members of the Temple of Mysteries, a brotherhood dedicated to the pursuit of life's secrets. An aspirant to the level of the Third Grade of the Arcanum, the highest rank in the fellowship's hierarchy, Marcus was preparing to take part in an unimaginably daring experiment. He was about to transport his mind beyond the point of death and then return it to the world of the living. The concoction he had readied for his bold voyage was a powerful mixture of consciousness-altering drugs. Marcus had tried some of the ingredients on a dog and they had seemed to induce the desired effects. But nobody knew what the full mixture would yield. Fearing he might never see his friend again, Clinias reflected that some truths should perhaps remain hidden from us. Stifling his doubts, he closed the heavy oak doors and left the sanctum without looking back at his friend.

Almost an hour had passed when Clinias was suddenly jolted by a haunting wail coming from the inner sanctum. He rushed to his friend's help and found him standing there, aghast. The secret of death had been revealed to Marcus, and the vision horrified him. "Death is reversal . . ." Marcus muttered. First comes reversal of consciousness, he realized. Before death, we experience the world from the inside out. After we die, consciousness remains, but it becomes external to the body. We see everything, including ourselves, from a purely objective perspective. We see our dead body lying there, being buried, and then slowly rotting. There is no escape from this predicament because consciousness is always located with our body. After a while comes the second reversal. Time begins to flow backward, from death back to life, from the end of life back to its beginning. With the shock of birth, consciousness and time reverse once again. The cycle never ends; it is never broken. The unfathomable truth about the human condition is that our souls are doomed to eternally oscillate between the poles of birth and death. Marcus knew all this now, but he wished that he didn't. He longed for a way to gain oblivion or a new life, but he did not know how to achieve such a thing.

This is my summary of another story by Barrington Bayley, called *Life Trap*.[1] Our soul-searching journey began with the tale of Harry and Rodrick, who faced the harrowing prospect of a soulless existence. In *The God Gun*, Bayley tells us that the soul is a blessing. This conclusion is also part of the dominant narrative in our culture today. By the end of his book, however, Bayley is claiming that this illusory blessing is in reality nothing but a thin layer of varnish masking a much deeper and darker truth. In *Life Trap*, the story of Clinias and Marcus, Bayley warns us that the soul might be a curse. In the previous chapter of this book, we saw that the ominous conclusion reached by Bayley in *The God Gun*, repeated ad nauseam by merchants of superstition in our culture, is nothing but a convenient fiction. As we discovered, we have nothing to lose by letting go of our soul beliefs.

In this final chapter, we are now poised to see soul beliefs for what they really are, and to reflect on Clinias and Marcus's deeper truth. Because soul beliefs inevitably distort our perception of reality, they corrupt our thinking on matters that are deeply important to all of us. By subverting the intuitions that underpin our criminal-justice system (i.e., influencing how we intuitively reason about moral responsibility), poisoning the debate over abortion, and muddling the question of whether people should have the right to die with dignity, soul beliefs, far from the blessing they are commonly portrayed to be, actually stand in the way of a healthier and more humane society. This is the soul's dark secret, the hidden truth that Barrington Bayley tried to warn us about in *Life Trap*. Like Clinias and Marcus, we too, denizens of twenty-first-century America, may be the ones who find ourselves trapped within the walls of unreason.

In a country in which religion is marketed as aggressively as fast food and prescription drugs, perhaps we shouldn't be too surprised to discover that the virtues of soul beliefs have been exaggerated by merchants of superstition who are also all too eager to cast "scientific materialism," their kryptonite, in the worst possible light, as we saw earlier. At the end of *Life after Death: The Evidence*, Dinesh D'Souza joins the chorus of conservative voices in America and warns his readers that "unbelief is neither intellectually plausible nor practically beneficial."[2] But under the light of reason, it is plain to see that D'Souza's claims are as hollow as the immaterial soul that he so desperately tries to defend. "Unbelief," as I have taken great pains to show you throughout this book, is precisely the attitude that reason and evidence compel us to adopt.

What about the charge that "unbelief" is not practically beneficial? After all, what good would knowing the truth be if it also made us miserable? In these final pages, I will show you that the ominous association of unbelief with unhappiness is nothing more than vulgar fearmongering. What's more, the New Dualists have it exactly backward. The empowering conclusion I want to share with you is that the truth about our material nature also contains the seeds of human flourishing. In denying the existence of the soul, I may have taken something away from you, but as we approach the end of our journey, I want to leave you with a reason to rejoice, secure in the knowledge that truth and happiness are not incompatible. I will show you that there is indeed grandeur in the naturalistic view of life—and death—and that we should embrace the conclusions we have reached throughout this book. The naturalistic worldview boldly asserts that we do not need to rely on ancient myths to find that existence has value and meaning. Yes, we are material and mortal creatures. This is what science teaches us. It also happens to be, as I will show you in the pages ahead, what gives our lives meaning and purpose.

CRIME AND PUNISHMENT

At one of the GOP debates during the run-up to the 2012 presidential election, NBC news anchor Brian Williams reminded candidate Rick Perry, then governor of Texas, that under his watch the Lone Star State had executed 234 death row inmates, a record high in modern times. Williams asked Perry whether he ever worried that some of those inmates might have been innocent. Undaunted, Perry replied that he never lost any sleep over this issue because the state of Texas had a very clear and thoughtful process in place. As Williams mentioned the number of people who were executed in Texas, the crowd erupted in cheers. Puzzled by this reaction, Williams asked Perry why he thought that the mention of the execution of 234 human beings drew applause from the crowd. "I think Americans understand justice," Perry replied.[3]

Perry's answer raises an interesting question. Where does our sense of justice come from and why exactly do we punish people? The answer to these questions is intimately related to our conception of human agency and to the question of free will and moral responsibility that we explored in the previous chapter. As we saw, the way people make sense of these issues hinges on whether they believe that human beings have immaterial souls and can therefore act as uncaused causes. If our dualistic intuitions are implicated in our understanding of responsibility, it might be worth asking to what extent these intuitions influence our conceptions of justice, and if they do, whether the consequences are desirable for our criminal-justice system.

In a provocative and widely cited article titled "For the Law, Neuroscience Changes Nothing and Everything," psychologists Joshua Greene and Jonathan Cohen argue that our intuitive sense of justice is indeed based on our intuitive dualism. Moreover, they conclude that this marriage of dualism and justice stands in the way of a more humane system of criminal justice, especially in a country like the United States.[4] For Greene and Cohen soul beliefs are a curse, just like they were to Clinias and Marcus in *Life Trap*. This is a particularly interesting argument, well worth examining in detail, because, if valid, it would turn the tables on the widespread assumption that soul beliefs are beneficial to people and society. If Greene and Cohen are on the right track, our intuitive soul beliefs are detrimental to our conception of criminal responsibility, and they should therefore be challenged on practical as well as moral grounds.

So why do we punish people? There are two standard responses to this question. According to the *consequentialist* approach, itself anchored in classical utilitarianism, the aim of punishment is forward-looking. We punish the guilty in order to contain dangerous individuals and prevent future crime through deterrence. To be sure, few would deny that crime prevention and the protection of the public are worthy pursuits. However, critics of consequentialism argue that this approach fails to capture the fundamental justification for punishment, *retribution*. This intuition is better captured, the argument goes, by the backward-looking, retributivist approach to punishment. On this view, we punish those who engage in criminal behavior because they *deserve* to be punished for their actions.

To illustrate these two approaches to punishment, let us revisit our example from chapter 8, the young man who killed the store clerk. For an advocate of consequentialism, the young man should be punished for his crime because of the danger he poses to society—he could strike again—and because punishing him will also send a message to would-be criminals: engage in unlawful behavior and you'll pay the price. Because consequentialism is concerned with the likely effects of punishment, and not with the question of whether someone is ultimately innocent or guilty in a deep metaphysical sense, this approach is compatible with different solutions to the problem of free will. Whether people's actions are fully caused or emanate from a contra-causal will is simply irrelevant. Consequentialism also meshes well with the law, which, as we saw, minimally assumes that people have a general capacity for rationality, a truism that is unlikely to be challenged by changing scientific explanations for this fundamental human capacity.

By contrast, for advocates of retributivism, the reason the young man should be punished is because he really *deserves* it. At the moment he was about to pull the trigger, our dualistic, contra-causal intuitions tell us, the young man *could have chosen to do otherwise*, but he nevertheless decided to open fire, and that makes him a "bad" person who deserves to be punished. Punishment on this view is not about containment or deterrence, although those are certainly desirable consequences; it is, above all, about giving those who misbehaved what they deserve. As Greene himself puts it, "For the retributivist, the ideal world is one in which the good people are rewarded and the bad people suffer."[5]

Unlike consequentialism, Greene and Cohen argue, retributivism is based on a much narrower view of human agency and a libertarian conception of free will. What the law cares about is whether the young man was sufficiently rational at the moment he pulled the trigger. But people care about something much deeper. What we want to know about the young man is whether it was really *him* who pulled the trigger. Was it *him* or his raging emotions? Was it *him* or his genes, his childhood, or his brain? This dualism between the "real person" and his physical self is what we've been talking about throughout this book. Greene and Cohen point out that even experts on criminal behavior and the brain can occasionally fall prey to dualism. They cite the following remarks by one such expert, Jonathan Pincus:

> When a composer conceives a symphony, the only way he or she can present it to the public is through an orchestra. . . . If the performance is poor, the fault could lie with the composer's conception, or the orchestra, or both. . . . Will is expressed by the brain. Violence can be the result of volition only, but if a brain is damaged, brain failure must be at least partly to blame.[6]

Pincus's analogy implies that there is a *you*, the composer, and another *you*, the orchestra, representing your brain. But as we saw in previous chapters, this dualistic distinction between you and your brain rests on an illusion.

Now here's the interesting part: If contra-causal free will is a fiction because the young man's actions were fully caused by factors ultimately beyond his control—his genes and environment, the chemistry of his brain, and so forth—then retributivism, which depends on a libertarian conception of free will, is flawed, however intuitively satisfying it may be. This argument was famously articulated by American lawyer and leader of the American Civil Liberties Union Clarence Darrow (1857–1938) in his 1924 defense of Nathan Leopold and Richard Loeb, two teenagers who were on trial for the kidnapping and murder of fourteen-year-old Bobby Franks, in what was known at the time as "the crime of the century." In his summation, Darrow pleaded for life in prison instead of death by hanging. Loeb and Leopold's actions, Darrow told the jury, resulted from forces over which the two young men had no control. "Nature is strong and she is pitiless. . . . We are her victims," he declared. "Each act, criminal or otherwise, follows a cause; given the same conditions, the same result will follow forever and ever."[7] In other words, the two teenagers *could not have chosen to do otherwise.*

By rejecting retributivism, Darrow certainly didn't suggest that Loeb and Leopold should be set free on the grounds that they were "victims of nature" and should therefore be exempt from responsibility. Consequentialism does entail that the two young men should be punished for their crime. But above all, Darrow believed, they should be treated the way all human beings should be treated, that is, humanely. Almost a century after Darrow's landmark defense of Loeb and Leopold, Greene and Cohen reiterate his plea: "At this time," they write, "the law deals firmly but mercifully with individuals whose behavior is obviously the product of forces that are ultimately beyond their control. . . . Someday," they add, "the law may treat all convicted criminals this way. That is, humanely."[8] In his book *Moral Tribes*, Greene develops this line of reasoning further and makes the case for a thoroughly consequentialist approach to moral responsibility and criminal justice.

In a country like the United States, the difference between consequentialism and retributivism can be a matter of life and death, quite literally. Americans may "understand justice," but human-rights advocates have long noticed that the United States remains the only advanced Western democracy that still imposes the death penalty on its citizens. A 2008 Gallup poll found that a solid majority of Americans continue to support capital punishment. According to the report, many Americans recognize that capital punishment is not a deterrent to crime, and even acknowledge that it has led to the execution of innocent people, but most nevertheless remain in favor of the death penalty as punishment for murder.[9]

The reason, the report concludes, is anchored in the conception of justice prevalent in the United States—Rick Perry may have been right after all. According to another Gallup poll, almost half of the respondents who were in favor of the death penalty invoked retribution as the main motivation, often mentioning the biblical injunction "eye for eye; tooth for tooth."[10] In the rest of the advanced Western democracies, capital punishment is regarded as a profound violation of human rights and an intolerable abuse of government power.

In its campaign to abolish the death penalty worldwide, the human-rights organization Amnesty International presents an arresting list of facts demonstrating that capital punishment does not work. According to a growing body of research, the death penalty is a costly practice that does not deter crime, it is biased against the poor and minorities, and it claims the lives of innocent people. Many organizations, including faith-based groups, victims' rights groups, national organizations, and state groups have reached the same conclusion. Amnesty International opposes the death penalty in all circumstances because it "violates the right to life and is the ultimate cruel, inhuman and degrading punishment."[11]

Sadly, the use of the death penalty in America only represents the tip of a much larger and more sinister iceberg. In a 2012 *Time* article titled "Incarceration Nation," journalist Fareed Zakaria describes what he calls "one of the great scandals of American life"—mass incarceration on a scale unprecedented in human history.[12] Commenting on this alarming state of affairs, *New Yorker*'s Adam Gopnick writes, "Overall, there are now more people under 'correctional supervision' in America—more than 6 million—than were in the Gulag Archipelago under Stalin at its height."[13]

In 2012, the United States had over 700 prisoners per 100,000 citizens, a rate up to ten times higher than in other developed countries. The number in Japan is 63 per 100,000, 90 in Germany, 96 in France, 97 in South Korea, and 153 in Britain. Even in developing countries like Mexico and Brazil, which are notorious for their crime problem, the numbers are a fraction of what they are in the United States (211 for Mexico and 274 for Brazil).

This trend toward mass incarceration is a recent development in American history that took place over the last two decades of the twentieth century. It is also in large part due to the failed "war on drugs." Back in 1980, the incarceration rate in the United States was 150 prisoners per 100,000 citizens, roughly the rate found in Britain today. Since then, the incarceration rate in the United States has quintupled. In his book *Thinking about Crime*, Michael Tonry, a professor of criminal law and policy at the University of Minnesota, describes the US criminal-justice system as "a punishment system that no one would knowingly have built from the ground up. It is often unjust, it is unduly severe, it is wasteful, and it does enormous damage to the lives of black Americans."[14] Tonry writes that the staggering increase in incarceration rates has little to do with actual crime rates and that, in general, punishment policies do not significantly reduce crime.

A historical analysis of crime and punishment trends in the United States, Germany, and Finland between 1960 and 1990 reveals that the evolution of crime rates in those three countries over that period was virtually identical. Homicide rates more than doubled, and violent crime rates grew by a factor of three to four. And yet, during the same period, the US incarceration rate quadrupled, whereas the German rate remained stable, and the rate in Finland fell by 60 percent. In the last two decades of the twentieth century, the difference in crime rates in Canada and the United States remained roughly constant, with the Canadian rate above the US rate. During the same period, the incarceration rate in Canada remained largely the same, while the US rate more than tripled, reaching unprecedented highs in 2002. Tonry points out that most experts and scholars no longer believe that severe crime-control policies substantially reduce crime.

The US criminal-justice system is highly retributive and it is also massively dysfunctional. The causes of this sad state of affairs involve the interaction of multiple factors, including the structure and organization of our system of governance, but Tonry also explains that "ordinary Americans make these things happen. Elected politicians proposed and enacted laws, but they would not have done it if they believe American voters would disapprove."[15]

This brings us back to Greene and Cohen's argument and Rick Perry's comment that Americans understand justice. If people's retributive impulses are ultimately anchored in their intuitive dualism, then soul beliefs are directly implicated in the failure of our criminal-justice system. Recall that in his global assessment of the relationship between religiosity and societal health, Gregory Paul concludes, "Data correlations show that in almost all regards the highly secular democracies consistently enjoy low rates of societal dysfunction, while pro-religious and antievolution America performs poorly."[16] If the argument developed by Greene and Cohen is on the right track, the unusually high levels of religiosity found in the United States are not merely correlated with societal dysfunction, they may very well, in the case of intuitive dualism, represent one of the causal factors. As far as justice is concerned, the soul might very well be a curse.

AT THE EDGES OF LIFE

During the 2012 presidential campaign, one of the billboards I drove by as I approached the City of Brotherly Love on my way home from New Jersey was a Republican Super PAC–sponsored ad that read: "Obama supports gay marriage and abortion. Do you?—Vote Republican." This billboard, still available on the Internet for those who suffer from campaign nostalgia, stands as a testament to the endless, and endlessly polarizing, debate in America over the question of abortion, passionately rehashed every election cycle. This is the eternal battle for the soul of the unborn. As you probably gathered from what I wrote, I grew up in France and I am a University Professor, so you can easily imagine where I stand on the question of abortion. However, my goal is not to convince you of my views on this hot-button issue. Nor will I claim that the issue can be easily settled. Instead, I want to focus on the contentious issue that lies at the heart of the abortion debate, and show you how religion and the question of the soul poison the debate.

So what's the big fight about? Life is a continuous progression, from sperm and egg cell to zygote, blastocyst, embryo, and then to fetus, and there is no fuss about the boundaries of this continuum. Nobody is seriously arguing that male masturbation should be made illegal because it is mass murder, or that it would be permissible to abort a healthy fetus toward the end of the third trimester. Somewhere between these two unambiguous boundaries, the line must be drawn. And this is where passions get stirred. To most people who oppose abortion, the line is drawn at the moment of conception, when they believe human life begins, and abortion is therefore unambiguously equated with murder. Most people who believe that abortion should be legal are not "pro-murder," of course. They simply disagree that the line should be drawn at the moment of conception. If we draw the line further down the developmental sequence, the argument goes, this gives women a window of opportunity during which they can exercise their freedom of choice.

In a 1984 debate between President Ronald Reagan and Walter Mondale, Jimmy Carter's vice president, Reagan cleverly framed the question by asserting that "until and unless someone can establish that the unborn child is not a living human being, then that child is already protected by the Constitution, which guarantees life, liberty, and the pursuit of happiness to all of us."[17] This is a clever argument, but it is utterly fallacious because it rests on a classic inversion of the logic underlying the burden of proof. For Reagan, the line is drawn at the moment of conception, and he challenges his would-be detractors to "prove him wrong." This is like the judge in our example from chapter 2 asking you to prove that you weren't speeding. As always, the burden of proof falls on the person making the claim. Someone may want to argue that the line must be drawn at the moment of conception, or at eight weeks later, or at some point during the second trimester, but in each case, he must provide evidence to support his decision to draw the line *at that particular point*. Reagan's decision to draw the line at the moment of conception doesn't win by default.

One of the virtues of the logic underlying the burden of proof is that it rests on a very democratic principle that doesn't only apply to advocates of the pro-life position. In *Moral Tribes*, Joshua Greene offers a utilitarian argument in support of the pro-choice position. For him, the moment of conception is not where the line should be drawn. But Greene doesn't challenge his readers to "prove him wrong." Instead, he offers a detailed argument showing that, all things considered, keeping abortion legal would probably maximize overall human happiness. Greene could, of course, be wrong, but at least he approaches the problem in a way that is intellectually honest. Likewise, Michael Shermer, in his book *The Science of Good and Evil*, defends the pro-choice position by arguing that abortion is a personal moral choice and should not be treated as a social or political issue. Like Greene, Shermer offers a detailed argument to justify his conclusion; he doesn't merely assume that he must be right until or unless someone can "prove him wrong."

All this shows that if pro-life advocates want to draw the line at the moment of conception and brand women who have abortions as murderers, they'd better have a good argument—in fact a *really* good one, given the nature of the accusations. Anyone accused of murder should be presumed innocent until proven guilty. No one is guilty by default, as Reagan's claim would entail. What is truly alarming given the stakes involved is that the debate over abortion is all too often framed not as a careful exercise in reason, as it should be, but as a religious or a theological question. In a September 2008 *New York Times* article, vice presidential hopeful Joe Biden, hardly the archetypal conservative, was quoted as saying that he is prepared "as a matter of faith to accept that life begins at the moment of conception."[18] On an ABC interview around the same period, then senator Barack Obama more modestly admitted that he doesn't "presume to be able to answer these kinds of theological questions."[19]

So, even at the more liberal end of the political spectrum, the issue of abortion is routinely framed in religious or theological terms. In an in-depth analysis of public opinion on abortion over several decades, the Gallup organization found that "the depth of one's religious beliefs, not gender, is what drives attitudes on abortion."[20] The analysis further revealed that most Americans (68 percent) who are "very religious" oppose abortion in almost all cases, but that most Americans (71 percent) who are "nonreli-gious" believe that abortion should be legal in all or most cases. Looking at people's attitudes within different faiths, the report concludes that the more fervently religious people in each faith also happen to be the ones who are the most inclined to the pro-life position compared to less-devout believers. Geographical patterns tell the same story. In the more-religious South and Midwest, people are more supportive of strict limits on abortion than are people in the less-religious East and West. A recent Pew analysis found that the regional divide between New England and the South-Central region over support for legal abortion has significantly deepened over the last twenty years.[21]

It is not difficult to see why religion plays such an important role in shaping people's attitudes on the question of abortion. Religion teaches us that life is a gift from God, who bestows upon each human being an immortal soul. In a 2008 educational article called "Do Embryos Have Souls?" the National Catholic Bioethics Center explains why abortion should be regarded as murder. The author of the article, Reverend Tadeusz Pachol-czyk, is a theologian with impeccable credentials: a doctorate in neuroscience from Yale followed by postdoctoral work at Harvard. Pacholczyk opens his commentary by observing that "People are sometimes surprised to hear that the wrongness of destroying a human embryo does not ultimately depend on when . . . he or she might receive a soul from God."[22] In other words, religiously inclined individuals usually believe that abortion is murder because God infuses embryos with a soul at the moment of conception.

The good reverend goes on to build an argument that is as clever as it is maddening. First, he points out that the point at which an embryo receives its soul, a process known as ensoulment, cannot be the issue because "the magisterium of the Church has never definitively stated when the ensoulment of the human embryo takes place." To be sure, the process of ensoulment has been debated for centuries within the Christian tradition, and the argument continues to this day. At the heart of the debate lies the question of whether ensoulment is *immediate* (whether it happens at the moment of conception) or *delayed* (whether it happens at a point after conception). It's easy to see why this is an important question for the church. Imagine that theologians were to unanimously decree that ensoulment is delayed. Would this not immediately open the door to advocates of the pro-choice position who could argue that soulless zygotes do not yet count as full human beings?

This is where the Catholic Church, according to our Ivy League–trained theologian, pulls its brilliant move and checkmates the pro-choice position. Destroying an ensouled embryo would certainly count as murder, but aborting an embryo that has not yet received its soul from God would be *even worse than murder*. This is because doing so would prevent the embryo from ever receiving a soul! And this would be terrible, far worse than murder, because the immortal soul is the vehicle through which the embryo would later be reunited with God in heaven. So, in the end, the church doesn't have to settle on when exactly embryos get ensouled. As Pacholczyk writes, "it is God's business as to precisely when He ensouls embryos." The crucial point for the Catholic Church is that we have to act *as if* embryos were ensouled at the moment of conception. Game, set, and match.

For religious opponents of abortion, the logic of ensoulment guarantees that the line must unambiguously be drawn at the moment of conception. Consequently, abortion is unequivocally equated with murder, if not worse. But as we have seen throughout this book, there is not a shred of respectable evidence for the existence of the soul. Believers in ensoulment, therefore, draw the line in the most capricious, arbitrary, and irrational way. That being said, I do not want to paint all religious people with the same brush. For some, the soul may not be implicated in the question of abortion. Nevertheless, most Americans are religious, and religious people oppose abortion to a much larger extent than do less-religious or nonreligious people. It should go without saying that implementing laws or policies intended to apply to all women on the basis of capricious decisions anchored in irrational beliefs is not a good idea. That in the twenty-first century an argument ultimately anchored in religious beliefs, soul-based or otherwise, can be used as a political wedge to brand women who have abortions as murderers is another scandal of American life.

In Barrington Bayley's fictional world, the soul imprisons us between the walls of birth and death. In twenty-first-century America, soul beliefs have an eerily similar effect. At one end of the continuum, they pollute the debate over abortion; at the other end, they close the door on the question of whether people should have the right to die with dignity. A little over ten years ago, obtaining the right to die became a desperate call for a young man from northern France, whose story ignited a national debate on this controversial question. On September 24, 2000, Vincent Humbert, a nineteen-year-old firefighter, was the victim of a tragic car accident that left him in a coma. When he awoke nine months later, Vincent found himself paralyzed, unable to utter a word, and practically blind. Only his intellectual abilities, his hearing, and minimal use of his right thumb were spared. Like the prisoners of silence we met in chapter 3, Vincent was trapped in a horrible nightmare. With the use of his right thumb, he was able select letters of the alphabet that were read aloud to him, and he could thus share the details of his tragic plight.

In a book called *I Ask for the Right to Die*, Vincent described his life as "a shit life. It is not a life, it is not my life. I can lead it no longer; I will lead it no longer."[23] In 2002, Vincent wrote to Jacques Chirac and implored the French president to grant him the right to die. "I would like you to know that you are my last chance,"[24] he told the head of state. Chirac refused. With no hope in sight, Vincent's mother, helped by a physician, decided to take the matter into her own hands and grant her son his final wish. Together, Marie Humbert and Dr. Chaussoy put an end to the young man's ordeal. Vincent died on September 26, 2003, shortly before the publication of his book. The case was brought to justice, but Vincent's mother and Dr. Chaussoy were not convicted of any wrongdoing, a decision that left Marie Humbert bitterly disappointed. She wanted to be convicted so that her trial could serve as a national platform from which she hoped to make the case for the legalization of assisted suicide in France.

On our side of the Atlantic, the question of assisted suicide brings to mind the name Jack Kevorkian, a euthanasia activist nicknamed "Dr. Death," who was convicted of second-degree murder in 1999 for the direct role he played in the death of a terminally ill patient. At the heart of stories like Vincent's and Jack Kevorkian's lies another hot-button question: Should assisted suicide be a legal option for those who wish to decide the moment of their death? As in the case of abortion, there are two ways to answer the question. One is to let our hearts be guided by the light of reason. Marshall all the arguments in favor of assisted suicide, all the arguments against such a proposition, and let rational debate take place.

The other approach is to invoke religious dogma, sacred books, and invisible entities like God and the soul. A Christian educational article on the question of assisted suicide reminds its readers that "God alone has the right to initiate and terminate life."[25] Various passages from the Bible are cited in support of this conclusion. Another document articulating the general Christian position on this question explains that human beings were created with the distinct purpose of being reunited with God in the afterlife. Moreover, readers are told, the process of dying is spiritually important because it brings people's souls closer to God, and it should therefore not be disrupted for fear that it would "interrupt the process of the spirit moving towards God."[26] Christianity is not unique in this regard. Most religions are against assisted suicide for very similar reasons.

The influence of religion on people's attitudes toward assisted suicide was measured in a 2013 Pew survey. The survey reveals that, overall, the American public is divided on the question of whether laws should be passed that allow physician-assisted suicide for the terminally ill. However, a breakdown of attitudes by religious affiliation reveals that the nonreligious are generally much more inclined to support such laws compared to religious individuals, although there is substantial variation between different religious groups. For example, 85 percent of respondents who were unaffiliated with any religion said they would support doctor-assisted-suicide laws for people who are in a great deal of pain with no hope of improvement, while only 42 percent of white evangelical Protestants and 42 percent of black Protestants said they would. Among white Catholics and white mainline Protestants these numbers are higher, but not as high as the numbers reported for those unaffiliated with any religion.[27]

As of the writing of this book, only four states have passed laws that legalize physician-assisted suicide: Montana, Oregon, Vermont, and Washington. Americans in other states will just have to suck it up and suffer with dignity so that their nonexistent souls can be properly reunited with God. Or else, if they can afford to do so, they always have the option to travel to Switzerland, where physician-assisted suicide has been legal since the 1940s. Far from the blessing they are usually portrayed to be, soul beliefs come at a societal price that we all have to pay. At the edges of life, the soul gets in the way of important questions about women's right to choose and our right to die with dignity. And so, like Clinias and Marcus, we are all trapped between the poles of birth and death.

THREE GIFTS

I must have been about nine or ten when I let go of my soul. Growing up in France, in a small village near the Swiss border, I was exposed to catechism through the local parish and had as an instructor a young nun who used to play the guitar and sing to us. She also told us about God and the soul, as well as heaven and hell, and we took notes with our special fountain pens, the ones we used to write down important lessons. We were encouraged to ask questions too. One day the young woman told us about God's omnipresence. "He is everywhere!" she beamed. I was puzzled by this idea and wondered how someone could be everywhere at once. So I raised my hand and asked whether God was in my pen too. She told me that he was. That day, on my way home, I remember thinking that God, the soul, heaven, and hell were just another story grown-ups had made up to compel children to be good and obedient. I also had a strong sense that, at least for me, they were unnecessary. I didn't need to be told that someone in the sky—or in my pen—was watching, to know I should be good. Being good just seemed like the right thing to do. I never went back to listen to the nun play her guitar and tell us about God and the fate of my soul. I had become soul-free.

I was too young at the time to reflect on the virtues of my newfound approach to life or to realize that science and reason were on my side. I simply no longer bothered with the myth, and it felt like the right thing to do. I now know that the practical benefits of the naturalistic approach to life come in the form of three wonderful gifts. The first gift stems from the conception of death—or rather, the state of *being dead*—that follows from the approach we have developed throughout this book. As I will show you, the naturalistic approach to death is comforting. The second gift comes from the understanding that we are mortal creatures and that our lives are, therefore, finite. This realization offers us the gift of meaningfulness. The third gift, perhaps the most important of all, is the gift of freedom that comes from adopting the principles of science and reason. When combined, these three gifts form the basis of an empowering alternative to the soul narrative that we will explore shortly—truth *and* happiness.

To begin thinking about the first gift, let us return to the Mortality Paradox. The first part of the paradox, you will recall, is that we human beings are uniquely aware of our own mortality. The second part is that consciousness cannot simulate the state of being dead, and so death, of which we are fully aware, is at the same time unimaginable to us. The soul narrative offers a nifty way around the Mortality Paradox. Soul believers accept the first leg of the paradox. We are all going to die—that's just a fact of life. It is the second part of the paradox that soul beliefs are supposed to alleviate. By assuring people that consciousness will never be extinguished, the soul narrative saves us from the agony of trying to imagine the unimaginable—what it is like to be dead.

However, if our materialistic hypothesis is true, the last thing in the world that we should worry about is being dead. If consciousness is extinguished at death, as materialism claims, then there is simply nothing like it is to be dead. Being dead wouldn't feel dark, cold, lonely, miserable, or anything. Since there would be no more consciousness to experience anything at all, being dead should be the least of our worries. The materialist philosopher Epicurus, whom we met in chapter 2, was one of the first thinkers to realize that "death is nothing to us" because "when we are, death has not come, and when death has come, we are not."[28] Woody Allen once quipped, "I am not afraid of death. I just don't want to be there when it happens."[29] Apparently, he didn't read Epicurus.

Even though consciousness cannot simulate its own nonexistence, there are a number of ways to think about what it would be like to be dead, even though we cannot really experience it. Philosopher Arthur Schopenhauer wrote, "Life is to be regarded as a loan from death, with sleep as the daily interest on this loan."[30] Dreamless sleep, or general anesthesia, states during which consciousness is temporarily "turned off," are the closest we have, in the world of the living, to what it would be like to be dead. A few years ago, I underwent two minor procedures for which I needed general anesthesia. Before the first procedure, I must admit that I was a little nervous. It had been more than thirty years since I had any kind of surgery. What would the experience be like as an adult? Would "falling asleep" or waking up be uncomfortable or even painful? What if the drugs only paralyzed me, but I remained fully conscious during the procedure?

Fortunately, everything went very well, and I was totally amazed that something as seemingly magical as consciousness can be safely turned off at will with the help of basic chemistry. Before my second procedure, I actually looked forward to the anesthesia. Emboldened by my first experience, I decided to experiment the second time around and try to resist the effects of the drugs. After all, if our will emanates from an immaterial soul that is not subject to the laws of nature, this should be doable. "Mind over matter!" as the New Age gurus are fond of reminding us. I remember hearing the anesthesiologist tell me that the drug had been administered. "Good," I thought, "let's see how tough my will is." I was able to count to maybe five, and the next thing I remember was waking up in a different room and being offered some apple juice. It was as though a magician watching the movie of my life had excised a few scenes, just as someone would use iMovie to edit a video clip. There is simply nothing like it is to be unconscious. If unconsciousness is a preview of death, then I am certainly not afraid of *being dead*—and neither should you.

The second gift is a direct consequence of the first one. If there is nothing like it is to be dead, then life has a terminal deadline. Therein lies the source of its meaning and value. This conclusion is beautifully captured by one of Jean de La Fontaine's famous fables, the story of the astrologer who stumbled into a well:

> To an astrologer who fell
> Plump to the bottom of a well,
> "Poor blockhead!" cried a passer-by,
> "Not see your feet, and read the sky?"
> . . .
> But I've digress'd. Return we now, bethinking
> Of our poor star-man, whom we left a drinking.
> Besides the folly of his lying trade,
> This man the type may well be made
> Of those who at chimeras stare
> When they should mind the things that are.[31]

The story of the astrologer who stumbled into a well is a timeless indictment of the folly of unreason. La Fontaine's astrologists, who try to predict human destiny by looking into the heavens, are our modern-day New Dualists. They are the merchants of superstition whose lying trade has us stare at chimeras instead of appreciating the things that really are. Indeed, what makes something precious is its scarcity. Time is valuable to us precisely because we do not have an infinite amount of it. With only a finite future ahead of us, we can make rational choices about how to spend our time in ways that are meaningful to us. This approach is explored in Phil Zimbardo and John Boyd's fascinating book *The Time Paradox*. Time, they argue, is our most precious commodity, much more valuable than money, because it is a finite quantity that can never be replenished. This, of course, follows from the fact that our lives have boundaries. Zimbardo and Boyd also show that our attitude toward time plays a significant role in determining our success and happiness.

Who hasn't heard someone, a colleague or a friend, say that he has made a conscious decision to spend less time working and more time picking up a new hobby or simply enjoying the company of his friends and family? "I'm much happier this way," he might tell us. When people like this imagine themselves at the end of their lives, they realize that having spent a little more time at the office, having earned a little more money, and having sat in a few more traffic jams, isn't going to make them any happier in the end. But they would regret immensely not having spent enough time with their loved ones or with themselves. For the more religiously inclined, the recognition that life's value comes from its finiteness is an idea that can even be found in the Bible. In Ecclesiastes, for example, the author reminds us that death is a terminal condition: "The dead know nothing; they have no more reward, and even the memory of them is lost."[32] His advice, then, is that we should make the most of our days: "Go thy way, eat thy bread with joy, and drink thy wine with a merry heart. . . . Live joyfully with the wife whom thou lovest all the days of thy vanity."[33]

In a similar vein, being reminded of the shortness and fragility of life can give us a renewed sense of its meaning and beauty. When I was in graduate school, I was involved in a car accident that could have been devastating. My car was totaled, but fortunately I escaped the accident unscathed. Over the next few days, I remember experiencing a newfound sense of meaning and purpose as I went through my regular daily activities. All of a sudden, simply being able to walk, go to the gym, talk to a friend, or type on my computer felt like an extraordinary gift to me. This experience is, in fact, quite common. In *Immortality*, Steven Cave cites the findings of psychiatrists who work with the terminally ill and report, even in patients diagnosed with cancer, an "enhanced sense of living . . . a vivid appreciation of the elemental facts of life . . . and deeper communication with loved ones."[34] Life is precious because it is short, fragile, and unpredictable. As Steve Jobs reminded a group of college graduates assembled to listen to his Stanford commencement speech, "Death is very likely the single best invention of life."[35]

The third gift from science and reason is the gift of intellectual freedom. "Take the risk of thinking for yourself—much more happiness, truth, and beauty will come to you that way."[36] Those were the words of the late Christopher Hitchens, one of the New Atheists. As the adjective *new* suggests, there were prominent nonbelievers before Hitchens and his cohort. During the age of free thought in American history, toward the end of the nineteenth century, one of the most celebrated nonbelievers, a true champion of intellectual freedom, was Robert Ingersoll (1833–1899), known as "The Great Agnostic." Ingersoll was a lawyer by trade, and he was also a charismatic orator whose public lectures drew large crowds wherever he went. Nowhere is the power and beauty of free thought more vividly expressed than in what has come to be known as Ingersoll's vow. I usually find long quotes tedious and try to avoid them as much as possible, but I will make an exception for this one. Every word and every sentence of Ingersoll's vow is worth reading, and reading again.

> When I became convinced that the universe was natural,
> That all the ghosts and gods were myths,
> There entered into my brain, into my soul, into every drop of my blood,
> The sense, the feeling, the joy of freedom.
> The walls of my prison crumbled and fell.
> The dungeon was flooded with light
> And all the bolts and bars and manacles turned to dust.
> I was no longer a servant, a serf, or a slave.
> There was for me no master in all the wide world, not even in infinite space.
> I was free to think.
> Free to express my thoughts,
> Free to live in my own ideal.
> Free to live for myself, and those I loved.
> Free to use all my faculties, all my senses.
> Free to spread imagination's wings,
> Free to investigate, to guess, and dream and hope.
> Free to judge and determine for myself.
> Free to reject all ignorant and cruel creeds,
> All the inspired books that savages have produced,
> And the barbarous legends of the past.
> Free from sanctified mistakes and "holy" lies.
> Free from the fear of eternal pain,
> Free from the winged monsters of the night.
> Free from devils, ghosts and gods.

For the first time I was free.

There were no prohibited places in all of the realm of thought.

No error, no space where fancy could not spread her painted wings.

No chains for my limbs.

No lashes for my back.

No flames for my flesh.

No Master's frown or threat,

No following in another's steps.

No need to bow or cringe or crawl, or utter lying words.

I was free; I stood erect and fearlessly, joyously faced all worlds.

My heart was filled with gratitude, with thankfulness,

And went out in love to all the heroes, the thinkers who gave their lives

For liberty of hand and brain,

For the freedom of labor and thought to those who fell

On the fierce fields of war.

To those who died in dungeons, bound in chains,

To those by fire consumed,

To all the wise, the good, the brave of every land

Whose thoughts and deeds have given freedom to the sons of men.

And then, I vowed to grasp the torch that they held, and hold it high,

That light might conquer darkness still.

A NEW NARRATIVE

According to proponents of the afterlife, soul beliefs serve two important functions. First, they allow us to resolve the Mortality Paradox. As Dinesh D'Souza intones, "Is death the end, or is there something more? This is the ultimate question . . . it is the issue that makes every other issue trivial."[37] In denying the finality of death, the soul narrative, we are told, infuses our lives with hope and meaning. This is what Rick Warren reminded his readers when he wrote that "if this life is all there is, there is no basis for any meaning, hope, purpose, or significance to life. . . . The logical end of such a life is despair."[38] We have now reached the conclusion that vendors of immortality, like the astrologers in La Fontaine's fable, are the ones staring at chimeras— dangerous mirages that will lead us all to stumble into the well.

The supreme irony for merchants of superstition is that science and reason contain within them the germs of an empowering alternative to the soul narrative. We began exploring the basis of this alternative conception by focusing on three gifts from science and reason. Let us now see how these gifts can be used to fulfill the promises of soul beliefs: solving the Mortality Paradox and giving our lives meaning and purpose. The first gift from science and reason, the realization that we should not fear the state of being dead, dissolves the Mortality Paradox by directly negating its second premise. Once we accept the conclusion that there is nothing like it is to be dead, the Mortality Paradox loses its raison d'être. All we are left with is the realization that our lives are finite. This realization is the second gift from science and reason, and it holds the answer to the question of purpose and meaning. Our lives are precious and meaningful precisely because they come to us with a final deadline.

I must confess that I have always been baffled by the notion that without belief in God or the soul, life would be entirely devoid of meaning and purpose. This makes about as much sense as the claim that without fennel and anchovies, people would starve to death. What about all the other foods? As for life, what about all the things that all of us, believers and nonbelievers alike, routinely derive pleasure, joy, and fulfillment from? What about love, friends, family, science, sports, the arts, and countless other activities and states of mind, not to mention the occasional hardship, that have given people's lives meaning and purpose since time immemorial? Is it really so hard to believe that an individual blessed with a robust health, a loving family, friends, a rewarding occupation, decent means, and time to engage in personal pursuits could find meaning and purpose in life without adding to her long list of acquaintances a few invisible friends? People like Rick Warren and Dinesh D'Souza, who seem to think that the answer to this question is no, strike me as seriously lacking imagination.

Unlike religion, science does not have the pretention to tell us what we should find meaningful. It just tells us that meaning doesn't necessarily come from above. Moreover, the gift of freedom, so beautifully captured by Robert Ingersoll, gives us free rein to decide for ourselves, with or without the help of sacred books, what should matter to us and how we should live our lives. In a country like the United States, where social inequality is endemic, freedom to think for ourselves and challenge authorities of any kind is our road to salvation. We can always try to pray our problems away or tell ourselves that they will all be redressed in the afterlife. But in addition to being misguided, this approach is also a recipe for avoiding our responsibilities here and now. As history testifies, it is within our collective power to bring about meaningful change. Slaves, people of color, women, and gays and lesbians, did not have to wait for their souls to be rescued by God. Their lives were made better not by the promise of eternal bliss but by other men and women, believers and nonbelievers, who took it upon themselves to act in the present. Death is very likely the single best invention of life, Steve Jobs told the Stanford class of 2005. He added that this is because death is the agent of change.[39]

Thus, the realization that our days are numbered compels us to try to live in the present. Renewed focus on the here and now also encourages us to develop gratitude. It is often said that happiness is not having what you want but wanting what you have. The conclusion that we are creatures of bones and flesh frees us from the burden of obsessing about the fate of our souls, and inhibiting our selfish impulses allows us to be mindful of the plight of others. Not fearing death, rejecting immortality as a futile illusion, and developing attitudes that foster living in the present, selflessness, and gratitude, form the backbone of a tradition known as the *wisdom literature*. These ideas have old and venerable roots that can be traced back to the Epicureans and the Stoics in ancient Greece, and can also be found in parts of the Old Testament such as Ecclesiastes, Job, Proverbs, Psalms, and the Song of Songs. These ideas will also be instantaneously recognizable to anyone who practices yoga today or follows the teachings of Buddhism.

This is what Steven Cave calls the *wisdom narrative* in his fascinating book on immortality. Wisdom is what we have found at the end of our soul-searching journey. And yet, it was there from the beginning. For the believer, the soul offers a comprehensive worldview, tying together a range of deeply important ideas about what life is, when it begins and ends, where meaning comes from, and what makes us do the things that we do. Let go of the soul, and the entire edifice comes crashing down. But this act of apparent destruction is in fact part of a larger and much more meaningful act of creation. As finite, physical, and mortal creatures, we are not infallible, and there are times when we need to acknowledge our own mistakes and correct them. Only through this process of self-correction and improvement—two of the hallmarks of scientific thinking—can progress on a large scale truly take place.

As things stand, the idea of the soul may offer some comfort to those who cannot bear the idea of one day no longer existing. But this illusory comfort rests on false hopes and comes at an enormous societal price that we all have to pay. Far from being the positive idea that it is often portrayed to be, the notion of the soul is actually what stands in the way of progress and a more just society, one in which needless anguish and suffering could be avoided. And so it is time to put the soul to rest. It wasn't immortal after all. It enjoyed a long and prosperous life of many millennia, but like many other ideas, it fell victim to progress and eventually decided to gracefully retire, passing its bright torch to its lifelong physical companion, our beloved body. The good news is that, like the emperor's new clothes in Andersen's famous tale, our invisible, immaterial soul never existed in the first place. So unlike the emperor who found himself naked in front of a perplexed crowd, we remain fully clothed and can maintain our dignity. In finally freeing ourselves from the shackles of superstition and renewing Ingersoll's vow, we can look at the future with hope and confidence. The new worldview that replaces the old one is indeed better and wiser—you can bet your life on it.

NOTES

FOREWORD

1. N. K. Sandars, *The Epic of Gilgamesh: An English Version with an Introduction* (London: Penguin Books, 1972), p. 102.

2. National Academy of Sciences, *Teaching about Evolution and the Nature of Science* (Washington, DC: National Academy, 1998), p. 58.

3. John F. Haught, *God and the New Atheism: A Critical Response to Dawkins, Harris, and Hitchens* (Louisville, KY: Westminster John Knox, 2008), p. 18.

4. Chris Mooney and Sheril Kirshenbaum, *Unscientific America: How Scientific Illiteracy Threatens Our Future* (New York: Basic Books, 2009), p. 105.

5. Pew Research Center Poll, "Scientific Achievements Less Prominent Than a Decade Ago," 2009, http://people-press.org/files/legacy-pdf/528.pdf (accessed June 9, 2014).

6. Victor J. Stenger, *God: The Failed Hypothesis; How Science Shows That God Does Not Exist* (Amherst, NY: Prometheus Books, 2007).

7. Harris Poll, "Americans' Belief in God, Miracles and Heaven Declines," 2013, http://www.harrisinteractive.com/NewsRoom/HarrisPolls/tabid/447/ctl/ReadCustom percent20Default/mid/1508/ArticleId/1353/Default.aspx (accessed June 9, 2014).

8. B. Libet et al., "Time of Conscious Intention to Act in Relation to Onset of Cerebral Activity (Readiness-Potential). The Unconscious Initiation of a Freely Voluntary Act," *Brain* 106 (1983): 623–42.

9. Alfred R. Mele, "Libet on Free Will: Readiness Potentials, Decisions, and Awareness," in *Conscious Will and Responsibility*, ed. Walter Sinnott-Armstrong and Lynn Nadel (Oxford; New York: Oxford University Press, 2011), pp. 23–33.

10. Chun Siong Soon et al., "Unconscious Determinants of Free Decisions in the Human Brain," *Nature Neuroscience* 11, no. 5 (2008): 545.

11. Victor J. Stenger, "Free Will and Autonomous Will: A Physicist's Perspective on How We Are Accountable for Our Actions," *Skeptic* 17, no. 4 (2012): 5–19.

12. Gregory S. Paul, "Cross-National Correlations of Quantifiable Societal Health with Popular Religiosity and Secularism in the Prosperous Democracies: A First Look," *Journal of Religion & Society* 7 (2005): 1–17.

13. Pippa Norris and Ronald Inglehart, *Sacred and Secular: Religion and Politics Worldwide*, 2nd ed. (Cambridge: Cambridge University Press, 2011).

14. Phil Zuckerman, *Society without God: What the Least Religious Nations Can Tell Us about Contentment* (New York: New York University Press, 2008).

CHAPTER 1: LIFTING THE VEIL

1. Barrington J. Bayley, *The God Gun*, in *The Seed of Evil* (London: Allison & Busby, 1979).

2. Mark C. Baker and Stewart Goetz, eds. *The Soul Hypothesis: Investigations into the Existence of the Soul* (New York: Continuum, 2011).

3. Pope John Paul II, "Magisterium Is Concerned with Question of Evolution: For It Involves Conception of Man," *Catholic Information Network*, October 22, 1996, http://www.cin.org/jp2evolu.html (accessed May 22, 2014).

4. Francis Crick, *The Astonishing Hypothesis: The Scientific Search for the Soul* (New York: Charles Scribner's Sons, 1994).

5. Owen Flanagan, *The Problem of the Soul: Two Visions of the Mind and How to Reconcile Them* (New York: Basic Books, 2002).

6. Joshua D. Greene, "Social Neuroscience and the Soul's Last Stand," in *Social Neuroscience: Toward Understanding the Underpinnings of the Social Mind*, ed. A. Todorov, S. Fiske, and D. Prentice (New York: Oxford University Press, 2011), pp. 263–73.

7. Flanagan, *Problem of the Soul*.

8. Peter Baker and Peter Slevin, "Bush Remarks on 'Intelligent Design' Theory Fuel Debate," *Washington Post*, August 3, 2005.

9. Paul Bloom, "Worse Than Creationism: Evolution, Neuroscience, and the Responsibility of Psychologists," *American Psychological Society: Observer* 18, no. 10 (2005), http://www.psychologicalscience.org/index.php/publications/observer/2005/october-05/worse-than-creationism-evolution-neuroscience-and-the-responsibility-of-psychologists.html. Physicist Victor J. Stenger echoed Bloom's remarks in a recent piece published in the *Huffington Post* titled "Science Is Politics," http://www.huffingtonpost.com/victor-stenger/science-is politics_b_53997.24html (accessed June 8, 2014).

10. Dinesh D'Souza, *Life after Death: The Evidence* (Washington, DC: Regnery, 2009), p. 13.

11. Joel Green, ed., *What about the Soul? Neuroscience and Christian Anthropology* (Nashville: Abington, 2004), back cover copy.

12. Richard Feynman, *Surely, You're Joking Mr. Feynman* (New York: W. W. Norton, 1985), p. 343.

CHAPTER 2: THE SPIRIT OF THE AGE

1. James P. Allen, ed., *Edwin Smith Surgical Papyrus,* National Library of Medicine, http://archive.nlm.nih.gov/proj/ttp/flash/smith/smith.html (accessed May 22, 2014).

2. This question was originally raised by the French neuroscientist Jean-Pierre Changeux in his superb book *Neuronal Man* (Princeton, NJ: Princeton University Press, 1997).

3. Andre Dollinger, "Body and Soul: An Introduction to the History and Culture of Pharaonic Egypt," September 2003, http://www.reshafim.org.il/ad/egypt/religion/body_and_soul.htm (accessed May 22, 2014).

4. Arthur C. Clarke, "Hazards of Prophecy: The Failure of Imagination," chapter 2 in *Profiles of the Future: An Enquiry into the Limits of the Possible* (New York: Harper & Row, 1962).

5. In their book, *A Brief History of the Soul* (Hoboken, NJ: Wiley-Blackwell, 2001), Stewart Goetz and Charles Taliaferro take issue with this conclusion (p. 155). They claim that for those in the Plato-Augustine-Descartes tradition, the existence of the soul has been affirmed mostly through what people are aware of from their first-person perspective. However, Goetz and Taliaferro's observation is only part of the story. Basing our judgment on a broader historical perspective, it is difficult to deny that our ancestors, including Plato and Descartes, were trying to account for life, mind, willed action, and consciousness when they invoked souls.

6. Jerome W. Elbert, *Are Souls Real?* (Amherst, NY: Prometheus Books, 2000), p. 36.

7. Stewart Goetz and Charles Taliaferro present a detailed and fairly objective overview of the history of soul beliefs in the first half of *A Brief History of the Soul* (see note 5). However, the second half of their book offers a very subjective and defensive interpretation of the modern clash between science and dualism.

8. Plato's proposed tripartite structure does not necessarily entail that he believed the soul was divisible into three parts. In fact, in *Phaedo*, Plato makes it clear that he takes the soul to be indestructible, imperishable, and without parts. As Goetz and Taliaferro suggest, Plato may be referring to three capacities of a single, indivisible soul, rather than to three separate souls, when he writes about the appetitive, spirited, and rational souls.

9. There is one passage in Aristotle's *De Anima*, cited on p. 27 of Goetz and Taliaferro's *A Brief History of the Soul*, which may suggest otherwise. Specifically, Aristotle reflects that the kind of soul associated with our intellect and contemplative faculty may indeed be separable—and therefore eternal—unlike the nutritive and sensitive souls which are not separable from the body.

10. A. C. Grayling, *Descartes: The Life and Times of a Genius* (New York: Walker, 2006).

11. René Descartes, *The World and Other Writings*, ed. and trans. S. Gaukroger (Cambridge: Cambridge University Press, 1998), p.169.

12. Ibid.

13. Stephen D. Snobelen, "Isaac Newton, Heretic: The Strategies of a Nicodemite," *British Journal for the History of Science* 32, no. 4 (1999): 381–419. As Snobelen shows, there are in fact reasons to doubt that Newton believed in the immortality of the soul.

14. As the great man himself put it: "I cannot conceive of a God who rewards and punishes his creatures, or has a will of the type of which we are conscious in ourselves. An individual who should survive his physical death is also beyond my comprehension, nor do I wish it otherwise; such notions are for the fears or absurd egoism of feeble souls." Albert Einstein, *The World as I See It* (Secaucus, NJ: Citadel, 1999), p. 5.

15. Neil deGrasse Tyson, "The Perimeter of Ignorance: A Boundary Where Scientists Face a Choice: Invoke a Deity or Continue the Quest for Knowledge," *Natural History Magazine* 28, no. 5 (November 2005); also available at Neil deGrasse Tyson, http://www.haydenplanetarium.org/tyson/read/2005/11/01/the-perimeter -of-ignorance (accessed September 16, 2014).

16. Barna Research Group, "Americans Describe Their Views about Life after Death," Barna Group: Knowledge to Navigate a Changing World, October 2003, https://www.barna.org/barna-update/article/5-barna-update/128-americans -describe-their-views-about-life-after-death#.UsWAGZg5v0c (accessed May 22, 2014).

17. Albert L. Winseman, "Eternal Destinations: Americans Believe in

Heaven, Hell," Gallup, May 2004, http://www.gallup.com/poll/11770/Eternal-Destinations-Americans-Believe-Heaven-Hell.aspx (accessed May 22, 2014).

18. Pew Research Religion and Public Life Project, "Chapter 1: Religious Beliefs and Practices," *U.S. Religious Landscape Survey: Religious Beliefs and Practices,* June 2008, http://www.pewforum.org/2008/06/01/chapter-1-religious-beliefs-and-practices/ (accessed May 22, 2014).

19. Larry Shannon-Missal, "Americans' Belief in God, Miracles and Heaven Declines," Harris: A Nielsen Company, December 2013, http://www.harris interactive.com/NewsRoom/HarrisPolls/tabid/447/ctl/ReadCustom%20 Default/mid/1508/ArticleId/1353/Default.aspx (accessed May 22, 2014).

20. Ibid.

21. Pippa Norris and Ronald Inglehart, *Sacred and Secular: Religion and Politics Worldwide* (New York: Cambridge University Press, 2004).

22. Steven Cave, *Immortality: The Quest to Live Forever and How It Drives Civilization* (New York: Crown, 2012), Kindle edition.

23. Douglas Hofstadter, *I Am a Strange Loop* (New York: Basic Books, 2008), p. 9.

24. Stephen T. Asma, "Soul Talk," *Chronicle of Higher Education,* May 2, 2010, http://chronicle.com/article/Soul-Talk/65278/ (accessed May 22, 2014).

25. Charles Francis Hockett, *Man's Place in Nature* (New York: McGraw-Hill, 1973), p. 133.

26. Edward J. Larson and Larry Witham, "Leading Scientists Still Reject God," *Nature* 394, no. 6691 (1998): 313, http://www.stephenjaygould.org/ctrl/ news/file002.html (accessed May 22, 2014).

27. Daniel Dennett, *Consciousness Explained* (Boston: Little, Brown, 1991), p. 33.

28. Joel Green, ed., *What about the Soul? Neuroscience and Christian Anthropology* (Nashville: Abington, 2004), back cover.

29. Steve Benen, "Political Animal," *Washington Monthly,* April 9, 2011, http://www.washingtonmonthly.com/archives/individual/2011_04/028869 .php (accessed May 22, 2014). For Steven Colbert's hilarious take on the incident, visit http://www.colbertnation.com/the-colbert-report-videos/381484/april -12-2011/jon-kyl-tweets-not-intended-to-be-factual-statements (accessed May 22, 2014).

30. Bertrand Russell, *Religion and Science* (Oxford: Oxford University Press, 1997), p. 137.

31. This is what Newton himself had to say about the force of gravity: "That

gravity should be innate, inherent and essential to matter, so that one body may act upon another at a distance thro' a vacuum, without the mediation of any thing else, by and through which their action and force may be conveyed from one to another, is to me so great an absurdity that I believe no man who has in philosophical matters a competent faculty of thinking can ever fall into it. Gravity must be caused by an Agent acting constantly according to certain laws; but whether this agent be material or immaterial, I have left to the consideration of my readers," Isaac Newton, *Letters to Bentley*, 1692/3.

32. Frank Newport, "In U.S., 46% Hold Creationist View of Human Origins," Gallup Poll, June 2012, http://www.gallup.com/poll/155003/Hold -Creationist-View-Human-Origins.aspx (accessed May 22, 2014).

33. Mark C. Baker and Stewart Goetz, eds., *The Soul Hypothesis: Investigations into the Existence of the Soul* (New York: Continuum, 2011), p. 5.

34. See note 31.

35. Noam Chomsky, *Language and Problems of Knowledge* (Cambridge, MA: MIT Press, 1988), p. 144.

CHAPTER 3: THE FIRST PRINCIPLE

1. Jon Palfreman, "Prisoners of Silence," PBS Frontline, October 19, 1993, http://www.pbs.org/wgbh/pages/frontline/programs/transcripts/1202.html (accessed May 23, 2014).

2. Ibid.

3. Ibid.

4. Oskar Pfungst, *Clever Hans (The Horse of Mr. Von Osten): A Contribution to Experimental Animal and Human Psychology* (New York: Henry Holt, 1911).

5. John Prichard, *A Few Sober Words of Table-Talk about Table-Spirits and the Rev. N.S. Godfrey's Incantations* (London: Simpkin and Marshall, 1853).

6. Uri Shwed and Peter Bearman, "The Temporal Structure of Scientific Consensus Formation," *American Sociological Review* 75, no. 6 (2010): 817–40.

7. James Alcock, "Back from the Future: Parapsychology and the Bem Affair," *Skeptical Inquirer*, March/April 2011; available at Committee for Skeptical Inquiry, http://www.csicop.org/specialarticles/show/back_from_the_future.

8. Jeff Galak et al., "Correcting the Past: Failures to Replicate Psi," *Journal of Personality and Social Psychology* 103, no.6 (2012): 933–48.

9. Michio Kaku, *Physics of the Impossible: A Scientific Exploration into the World of Phasers, Force Fields, Teleportation, and Time Travel* (New York: Anchor Books, 2009), p. 283.

10. Joseph P. Simmons, Leif D. Nelson, and Uri Simonsohn, "False-Positive Psychology: Undisclosed Flexibility in Data Collection and Analysis Allows Presenting Anything as Significant," *Psychological Science* 22, no. 11 (2011): 1359–66.

11. Richard Dawkins, interview by Bill O'Reilly, *O'Reilly Factor*, Fox News, posted on YouTube (5:10) by "Thiago Ramos" on September 30, 2011, https://www.youtube.com/watch?v=ULv2B51HY80 (accessed September 11, 2014).

12. Michael S. Gazzaniga, "Cerebral Specialization and Interhemispheric Communication: Does the Corpus Callosum Enable the Human Condition?" *Brain* 123, no. 7 (2000): 1293–326.

13. Michael S. Gazzaniga, "Neurological Disorders and the Structure of Human Consciousness," *Trends in Cognitive Sciences* 7, no. 4 (2003): 161–65.

14. Ibid.

15. Stanley Schachter and Jerome E. Singer, "Cognitive, Social, and Physiological Determinants of Emotional State," *Psychological Review* 69 (1962): 379–99.

16. Donald D. Hoffman and Manish Singh, "Computational Evolutionary Perception," *Perception* 41 (2012): 1073-91.

CHAPTER 4: DUALISM ON TRIAL

1. Duncan MacDougall, "Hypothesis Concerning Soul-Substance Together with Experimental Evidence of the Existence of Such Substance," *American Medicine*, April 1907, 240–43.

2. Stewart Goetz and Charles Taliaferro, *A Brief History of the Soul* (Hoboken, NJ: Wiley-Blackwell, 2011), p. 181.

3. Mark C. Baker and Stewart Goetz, eds., *The Soul Hypothesis: Investigations into the Existence of the Soul* (New York: Continuum, 2011), p. 101.

4. Stephen Colbert, "Speech at the White House Correspondents' Dinner" (Washington, DC, April 29, 2006), Daily Kos, http://www.dailykos.com/story/2013/04/27/1205314/-Colbert-s-Speech-at-the-White-House-Correspondents-Dinner# (accessed May 27, 2014).

5. Paul M. Churchland, *Matter and Consciousness* (Cambridge, MA: MIT Press, 2013), pp. 24–25.

6. Charles R. Gallistel, "Behaviorism, Methodological and Theoretical," in *Routledge Encyclopedia of Philosophy*, ed. Edward Craig (London: Routledge, 1998), pp. 696–99.

7. Stanislas Dehaene, *Consciousness and the Brain: Deciphering How the Brain Codes Our Thoughts* (New York: Viking, 2014), Kindle edition, chapter 1.

8. Ibid.

9. James Mauro, "Bright Lights, Big Mystery," *Psychology Today*, July 1992, http://www.psychologytoday.com/articles/200910/bright-lights-big-mystery (accessed September 11, 2014).

10. S. M. Simpson, "Near Death Experience: A Concept Analysis as Applied to Nursing," *Journal of Advanced Nursing* 36, no. 4 (2001): 520–26.

11. Mario Beauregard and Denyse O'Leary, *The Spiritual Brain: A Neuroscientist's Case for the Existence of the Soul* (New York: HarperOne, 2007), p. 166.

12. Dinesh D'Souza, *Life after Death: The Evidence* (Washington, DC: Regnery, 2009), p. 72.

13. Hayden Ebbern, Sean Mulligan, and Barry L. Beyerstein, "Maria's Near Death Experience: Waiting for the Other Shoe to Drop," *Skeptical Inquirer* 20, no. 4 (1996): 46–60.

14. Kenneth Ring and Sharon Cooper, *Mindsight: Near-Death and Out-of-Body Experiences in the Blind* (Palo Alto, CA: William James Center for Consciousness Studies, Institute of Transpersonal Psychology, 1999).

15. Victor J. Stenger, *God and the Folly of Faith: The Incompatibility of Science and Religion* (Amherst, NY: Prometheus Books, 2012).

16. Olaf Blanke, Stéphanie Ortigue, Theodor Landis, and Margitta Seeck, "Stimulating Illusory Own-Body Perceptions," *Nature* 419 (2002): 269–70.

17. Eben Alexander, *Proof of Heaven: A Neurosurgeon's Journey into the Afterlife* (New York: Simon & Schuster, 2012), p. 188.

18. Sam Harris, "Sam Harris Won't Debate Dr. Eben Alexander on Near-Death Experience Science," posted October 16, 2012, http://www.skeptiko.com/sam-harris-wont-debate-eben-alexander-on-near-death-experience-science/ (accessed May 27, 2014).

19. David Silverman, "Bill O'Reilly vs. David Silverman—Tide Goes In, Tide Goes Out," YouTube video, 5:26, Fox News, January 4, 2011, posted by "Tide Goes In, Tide Goes Out," February 17, 2011, https://www.youtube.com/watch?v=wb3AFMe2OQY (accessed May 27, 2014).

20. Bill O'Reilly, "Where Did It All Come From?" YouTube video, 1:43, BillOReilly.com, posted by "The Official BillOReilly.com Channel," January 26,

2011, https://www.youtube.com/watch?v=UyHzhtARf8M (accessed May 27, 2014).

21. Jean Bricmont, "Qu'est-ce que le Materialisme Scientifique?" Dogma, http://www.dogma.lu/txt/JB-MatSc.htm (accessed May 27, 2014).

22. D'Souza, *Life after Death*, p. 179.

23. Ibid., p. 168.

24. Ibid., p. 172.

25. Ibid., p. 182.

26. Ibid.

27. Ibid.

28. Ibid., p. 18.

29. Ibid., p. 15.

30. Baker and Goetz, *Soul Hypothesis*, pp. 73–74.

31. Ibid., p. 93.

32. Ibid., p. 91.

33. Ibid.

34. Alexander, *Proof of Heaven*, p. 153.

35. Ibid., p. 153.

36. Beauregard and O'Leary, *Spiritual Brain*, p. xii.

37. Richard Feynman, *The Character of Physical Law* (London: BBC, 1965), p. 129.

38. D'Souza, *Life after Death*, p. 74.

39. Victor J. Stenger, "Quantum Quackery," Committee for Skeptical Inquiry, January/February 1997, http://www.csicop.org/si/show/quantum_quackery/ (accessed May 27, 2014).

40. Leonard Mlodinow, "Deepak Chopra Faces a Real Theoretical Physicist," November 3, 2013, https://www.youtube.com/watch?v=0qFGs-SIWB4 (accessed May 27, 2014).

41. Maximilian Schlosshauer, Johannes Koffler, and Anton Zeilinger, "The Interpretation of Quantum Mechanics: From Disagreement to Consensus?" *Annalen der Physik* 525, no. 4 (2013): A51.

42. Ibid.

CHAPTER 5: REQUIEM FOR THE SOUL

1. Thomas Henry Huxley, *Essays upon Some Controverted Questions* (New York: D. Appleton, 1893), p. 268.

2. David Hume, *An Enquiry Concerning Human Understanding*, 1748 (Indianapolis: Hackett, 1993), p. 73.

3. Paul M. Churchland, *Matter and Consciousness* (Cambridge, MA: MIT Press, 2013), p. 35.

4. Bertrand Russell, *Religion and Science* (Oxford: Oxford University Press, 1997), p. 200.

5. Elizabeth of Bohemia, *Letter to Descartes*, May 6, 1643.

6. Peter Hoffman, *Life's Ratchet: How Molecular Machines Extract Order from Chaos* (New York: Basic Books, 2012), Kindle edition, chapter 3.

7. Jerry Fodor, "The Mind-Body Problem," in *The Mind-Body Problem*, ed. R. Warner and T. Szubka (Oxford: Blackwell, 1994), p. 25.

8. James C. Maxwell, *Theory of Heat*, rev. ed., ed. John William Strutt Rayleigh (London: Longmans, Green, 1902), p. 338.

9. Harold J. Morowitz, "The Mind Body Problem and the Second Law of Thermodynamics," *Biology and Philosophy* 2 (1987): 274.

10. Léon Brillouin, "Maxwell's Demon Cannot Operate: Information and Entropy I," *Journal of Applied Physics* 22 (1951): 334–37; and Léon Brillouin, *Science and Information Theory* (New York: Academic Press, 1956).

11. Morowitz, "Mind Body Problem," p. 275.

12. Dinesh D'Souza, *Life after Death: The Evidence* (Washington, DC: Regnery, 2009), p. 220.

13. Mark C. Baker and Stewart Goetz, eds., *The Soul Hypothesis: Investigations into the Existence of the Soul* (New York: Continuum, 2011), p. 12.

14. Victor J. Stenger, *The New Atheism: Taking a Stand for Science and Reason* (Amherst, NY: Prometheus Books, 2009), p. 186.

15. Stewart Goetz and Charles Taliaferro, *A Brief History of the Soul* (Hoboken, NJ: Wiley-Blackwell, 2011), p. 201.

16. Baker and Goetz, *Soul Hypothesis*, p. 125.

17. Ibid., pp. 11–12.

18. Ibid., p. 73.

19. Ibid., p. 252.

20. Ibid.

21. Christopher Hitchens, *God Is Not Great: How Religion Poisons Everything* (New York: Twelve Books, 2007), p. 150.

22. Baker and Goetz, *Soul Hypothesis*, p. 93.

CHAPTER 6: LA METTRIE'S REVENGE

1. Christopher Smith, "Julien Offray de La Mettrie (1709–1751)," *Journal of the History of the Neurosciences* 11, no. 2 (2002): 110–24.

2. Ibid.

3. A. C. Grayling, *Descartes: The Life and Times of a Genius* (New York: Walker, 2006).

4. Smith, "Julien Offray de La Mettrie."

5. Mark C. Baker and Stewart Goetz, eds., *The Soul Hypothesis: Investigations into the Existence of the Soul* (New York: Continuum, 2011), pp. 235–36.

6. Stewart Goetz and Charles Taliaferro, *A Brief History of the Soul* (Hoboken, NJ: Wiley-Blackwell, 2011).

7. Charles Sherrington, *The Integrative Action of the Nervous System* (New York: Charles Scribner's Sons, 1906), foreword to the 1947 edition, p. xviii.

8. Michael S. Gazzaniga, "Cerebral Specialization and Interhemispheric Communication: Does the Corpus Callosum Enable the Human Condition?" *Brain* 123, no. 7 (2000): 1293–326.

9. Todd E. Feinberg, *Altered Egos: How the Brain Creates the Self* (Oxford: Oxford University Press, 2001), Kindle edition.

10. Dinesh D'Souza, *Life after Death: The Evidence* (Washington, DC: Regnery, 2009), pp. 143–44.

11. I. Biran and A. Chatterjee, "Alien Hand Syndrome," *Archives of Neurology* 61 (2004): 292–94.

12. Feinberg, *Altered Egos*, Kindle edition, chapter 6.

13. H. Debruyne et al., "Cotard's Syndrome: A Review," *Current Psychiatry Reports* 11 (2009): 197–202.

14. Ibid.

15. For a great book on the dark side of the history of science, I recommend *Elephants on Acid and Other Bizarre Experiments* by Alex Boese.

16. Wilder Penfield, *The Mystery of the Mind: A Critical Study of Consciousness and the Human Brain* (Princeton, NJ: Princeton University Press, 1978), p. 77.

17. Stewart Goetz, "God and Mind/Body Dualism," *Reasonable Faith with William Lane Craig*, July 27, 2009, http://www.reasonablefaith.org/god-and-mind-body-dualism (accessed May 29, 2014).

18. Michel Desmurget et al., "Movement Intention after Parietal Cortex Stimulation in Humans," *Science* 324, no. 5928 (2009): 811–13.

19. A. T. Barker et al., "Non-Invasive Magnetic Stimulation of the Human Motor Cortex," *Lancet* 8437 (1985): 1106–1107.

20. E. Wassermann and J. Grafman, "Combining Transcranial Magnetic Stimulation and Neuroimaging to Map the Brain," *Trends in Cognitive Sciences* 1, no. 6 (1997): 199–200.

21. D'Souza, *Life after Death*, chapter 10.

22. Liane Young et al., "Disruption of the Right Temporoparietal Junction with Transcranial Magnetic Stimulation Reduces the Role of Beliefs in Moral Judgments," *Proceedings of the National Academy of Sciences of the United States of America* 107, no. 15 (2010): 6753–58.

23. This is the definition used by Steven Pinker in his book *How the Mind Works*.

24. Bruce Upbin, "IBM's Watson Gets Its First Piece of Business in Healthcare," *Forbes*, February 8, 2013, http://www.forbes.com/sites/bruceupbin/2013/02/08/ibms-watson-gets-its-first-piece-of-business-in-healthcare/ (accessed September 13, 2014).

25. J. V. Haxby, "Distributed and Overlapping Representations of Faces and Objects in Ventral Temporal Cortex," *Science* 293 (2009): 2425–30.

26. J. D. Haynes and R. Geraint, "Decoding Mental States from Brain Activity in Humans," *Nature Neuroscience* 7 (2006): 523–34.

27. Ibid.

28. Ibid.

29. Ibid.

30. Mario Beauregard and Denyse O'Leary, *The Spiritual Brain: A Neuroscientist's Case for the Existence of the Soul* (New York: HarperOne, 2007), p. xii.

31. In their glossary, *materialism* is defined as "the philosophy that matter is all that exists and everything has a material cause."

32. Beauregard and O'Leary, *Spiritual Brain*, p. xiii.

33. D'Souza, *Life after Death*, pp. 130–31.

34. Beauregard and O'Leary, *Spiritual Brain*, p. xiv.

35. D'Souza, *Life after Death*, pp. 114–15.

36. Stephen Cave, "What Science Says about the Soul," *Skeptic Magazine* 18, no. 2 (2013): 16–18.

37. D'Souza, *Life after Death*, p. 139.

CHAPTER 7: DESCARTES'S SHADOW

1. Sigmund Freud, "Thoughts for the Times on War and Death," in *The Standard Edition of the Complete Psychological Works of Sigmund Freud*, vol. 14 (London: Hogarth, 1915), 274–301.

2. Elizabeth Spelke et al., "Spatiotemporal Continuity, Smoothness of Motion and Object Identity in Infancy," *British Journal of Developmental Psychology* 13, no. 2 (1995):113–42.

3. Alan Slater, Victoria Morison, and David Rose, "Habituation in the Newborn," *Infant Behavior and Development* 7 (1984): 183–200.

4. Valerie Kuhlmeier, Paul Bloom, and Karen Wynn, "Do 5-Month-Old Infants See Humans as Material Objects?" *Cognition* 94 (2004): 95–103.

5. Amanda Woodward, "Infants Selectively Encode the Goal Object of an Actor's Reach," *Cognition* 69, no. 1 (1998): 1–34.

6. Kristine Onishi and Renée Baillargeon, "Do 15-Month-Old Infants Understand False Beliefs?" *Science* 308, no. 5719 (2005): 255–58.

7. Jesse Bering and David Bjorklund, "The Natural Emergence of Reasoning about the Afterlife as a Developmental Regularity," *Developmental Psychology* 40, no. 2 (2004): 217–33.

8. K. Banerjee, O. Haque, and E. Spelke, "Melting Lizards and Crying Mailboxes: Children's Preferential Recall of Minimally Counterintuitive Concepts," *Cognitive Science* 37, no.7 (2013): 1251–89.

9. Pew Research Global Attitudes Project, "Chapter 2: Religiosity," September 17, 2008, http://www.pewglobal.org/2008/09/17/chapter-2 -religiosity/ (accessed May 29, 2014).

10. Noam Chomsky, *The Prosperous Few and the Restless Many* (Berkeley, CA: Odonian, 1993), chapter 12; also available at http://zcomm.org/wp-content/ uploads/zbooks/htdocs/chomsky/pfrm/pfrm-12.html (accessed September 17, 2004).

11. Penny Edgell, Joseph Gerteis, and Douglas Hartman, "Atheists as 'Other': Moral Boundaries and Cultural Membership in American Society," *American Sociological Review* 71 (2006): 211–34.

12. "Joel Osteen: 'Homosexuality Is a Sin,'" YouTube video, 5:02, from CNN broadcast on January 25, 2011, posted by "Piers Morgan Tonight," January 25, 2011, https://www.youtube.com/watch?v=tgCpRNfBzys (accessed May 29, 2014).

13. "Rep. John Shimkus: God Decides When the 'Earth Will End,'" YouTube video, 2:25, posted by "Progress Illinois," March 25, 2009, https://www.youtube.com/watch?v=_7h08RDYA5E (accessed May 29, 2014).

14. Noam Chomsky, "On the Nature, Use, and Acquisition of Language," in *Handbook of Child Language*, ed. William C. Ritchie and Tej K. Bhatia (New York: Academic Press, 1999), p. 33.

15. Noam Chomsky, "Linguistics and Cognitive Science: Problems and Mysteries," in *The Chomskyan Turn*, ed. A. Kasher (Oxford: Basil Blackwell, 1991), pp. 26–53.

16. Daniel Dennett, *Intuition Pumps and Other Tools for Thinking* (New York: W. W. Norton, 2013), Kindle edition, chapter 7.

17. Dinesh D'Souza, *Life after Death: The Evidence* (Washington, DC: Regnery, 2009), p. 143.

18. Mark C. Baker and Stewart Goetz, eds., *The Soul Hypothesis: Investigations into the Existence of the Soul* (New York: Continuum, 2011), p. 9.

19. Mario Beauregard and Denyse O'Leary, *The Spiritual Brain: A Neuroscientist's Case for the Existence of the Soul* (New York: HarperOne, 2007), p. 104.

20. Eben Alexander, *Proof of Heaven: A Neurosurgeon's Journey into the Afterlife* (New York: Simon & Schuster, 2012), p. 154.

21. Adenauer G. Casali et al., "A Theoretically Based Index of Consciousness Independent of Sensory Processing and Behavior," *Science Translational Medicine* 5, no.198 (2013): 105.

22. Neil deGrasse Tyson, *The Sky Is Not the Limit: Adventures of an Urban Astrophysicist* (Amherst, NY: Prometheus Books, 2004), p. 38.

CHAPTER 8: THE SUM OF ALL FEARS

1. Penny Edgell, Joseph Gerteis, and Douglas Hartmann, "Atheists as 'Other': Moral Boundaries and Cultural Membership in American Society," *American Sociological Review* 71 (2006): 211–34.

2. Bob Herbert, "In America: The True Believer," *New York Times*, November 30, 2000, http://www.nytimes.com/2000/11/30/opinion/in-america-the-true-believer.html (accessed May 31, 2014).

3. Francis X. Clines, "Capitol Sketchbook: In a Bitter Culture War, an Ardent Call to Arms," *New York Times*, June 17, 1999, http://www.nytimes

.com/1999/06/17/us/capitol-sketch-
book-in-a-bitter-cultural-war-an-ardent
-call-to-arms.html?src=pm (accessed May 31, 2014).

4. Kel Kelly, "The Sandy Hook School Shootings Didn't Happen Because We Don't Allow God in Schools," *Huffington Post*, December 19, 2012, http://www.huffingtonpost.com/kel-kelly/sandy-hook-shooting_b_2308316.html (accessed September 19, 2014).

5. Mark C. Baker and Stewart Goetz, eds., *The Soul Hypothesis: Investigations into the Existence of the Soul* (New York: Continuum, 2011), p. 73.

6. Daniel Dennett should be credited for using this metaphor in his book *Freedom Evolves* to make a very similar point about free will.

7. "Libertarian Free Will," *Theopedia*, http://www.theopedia.com/Libertarian_free_will (accessed May 31, 2014).

8. Patricia Smith Churchland, "Moral Decision-Making and the Brain," in *Neuroethics: Defining the Issues in Theory, Practice, and Policy*, ed. Judy Illes (Oxford: Oxford University Press, 2006), pp. 3–16.

9. Eddy Nahmias, D. Justin Coates, and Trevor Kvaran, "Free Will, Moral Responsibility, and Mechanism: Experiments on Folk Intuitions," *Midwest Studies in Philosophy* 31 (2007): 214–42. See also Shaun Nichols, "The Folk Psychology of Free Will: Fits and Starts," *Mind & Language* 19, no. 5 (2004): 473–502.

10. David Hume, *An Enquiry Concerning Human Understanding*, 1748 (Indianapolis: Hackett, 1993).

11. Dinesh D'Souza, *Life after Death: The Evidence* (Washington, DC: Regnery, 2009), p. 139.

12. Ibid., pp. x–xi.

13. Stephen Cave, *Immortality: The Quest to Live Forever and How It Drives Civilization* (New York: Crown, 2012), p. 263.

14. Ibid., p. 161.

15. Matthew 22:23–30.

16. Phil Zuckerman, *Society without God: What the Least Religious Nations Can Tell Us about Contentment* (New York: New York University Press, 2008), p. 24.

17. Ibid., pp. 18, 19.

18. Ibid., p. 2.

19. Ibid., p. 66.

20. Ibid., p. 62.

21. Ibid., p. 4.

22. Ibid., p. 66.

23. D'Souza, *Life after Death*, p. 216.

24. Ibid., p. 199.

25. Ibid., p. 217.

26. Gregory S. Paul, "Cross-National Correlations of Quantifiable Social Health with Popular Religiosity and Secularism in the Prosperous Democracies: A First Look," *Journal of Religion & Society* 7 (2005): 1–17.

27. Ibid., p. 1.

28. Ibid., p. 7.

29. Ibid.

30. Paul Bloom, "Religion, Morality, Evolution," *Annual Review of Psychology* 63 (2012): 179–99.

31. Ibid., p. 190.

32. Robert Putnam and David Campbell, *American Grace: How Religion Divides and Unites Us* (New York: Simon & Schuster, 2012), p. 473.

CHAPTER 9: IMAGINE

1. Barrington J. Bayley, *Life Trap*, in *The Seed of Evil* (London: Allison & Busby, 1979).

2. Dinesh D'Souza, *Life after Death: The Evidence* (Washington, DC: Regnery, 2009), p. 217.

3. Rick Perry, "Rick Perry on Death Penalty and 'Ultimate Justice,'" YouTube video, 2:04, CNS News, September 8, 2011, http://www.youtube.com/watch?v=fXH7Z6M4vOs (accessed June 1, 2014).

4. Joshua Greene and Jonathan Cohen, "For the Law, Neuroscience Changes Nothing and Everything," *Philosophical Transactions of the Royal Society of London* 359 (2004): 1775–85.

5. Joshua D. Greene, *Moral Tribes: Emotion, Reason, and the Gap between Us and Them* (New York: Penguin, 2013), p. 269.

6. Jonathan Pincus, as quoted in Greene and Cohen, "For the Law," p. 1779.

7. Clarence Darrow, "Closing Argument," in *Famous American Trials: Illinois v. Nathan Leopold and Richard Loeb, 1924*, http://law2.umkc.edu/faculty/projects/ftrials/leoploeb/leopold.htm (accessed September 19, 2014).

8. Greene and Cohen, "For the Law," p. 1784.

9. Lydia Saad, "Americans Hold Firm to Support for Death Penalty," Gallup Poll, November 17, 2008, http://www.gallup.com/poll/111931/Americans

-Hold-Firm-Support-Death-Penalty.aspx (accessed June 1, 2014).

10. Jeffrey M. Jones, "Support for the Death Penalty 30 Years after the Supreme Court Ruling," Gallup Poll, June 30, 2006, http://www.gallup.com/poll/23548/Support-Death-Penalty-Years-After-Supreme-Court-Ruling.aspx (accessed June 1, 2014).

11. "Death Penalty 2012: Despite Setbacks, a Death Penalty-Free World Came Closer," Amnesty International, April 10, 2010, http://www.amnesty.org/en/news/death-penalty-2012-despite-setbacks-death-penalty-free-world-came-closer-2013-04-10-0 (accessed June 1, 2014).

12. Fareed Zakaria, "Incarceration Nation," *Time*, April 2, 2012, http://content.time.com/time/magazine/article/0,9171,2109777,00.html (accessed September 19, 2014).

13. Adam Gopnick, "The Caging of America: Why Do We Lock Up So Many People?" *New Yorker*, January 30, 2012, http://www.newyorker.com/magazine/2012/01/30/the-caging-of-america (accessed September 19, 2014).

14. Michael Tonry, *Thinking about Crime: Sense and Sensibility in American Penal Culture* (Oxford: Oxford University Press, 2004), Kindle edition.

15. Ibid.

16. Gregory S. Paul, "Cross-National Correlations of Quantifiable Societal Health with Popular Religiosity and Secularism in the Prosperous Democracies: A First Look," *Journal of Religion & Society* 7 (2005): 1–17.

17. "Debate between the President and Former Vice President Walter F. Mondale in Louisville, Kentucky," Ronald Reagan Presidential Foundation and Library, October 7, 1984, http://www.reaganfoundation.org/reagan-quotes-detail.aspx?tx=2041 (accessed June 1, 2014).

18. Kate Phillips, "As a Matter of Faith, Biden Says Life Begins at Conception," *New York Times*, September 7, 2008, http://www.nytimes.com/2008/09/08/us/politics/08campaign.html?_r=0 (accessed June 1, 2014).

19. Tahman Bradley, "Obama: My Answer on Abortion at Saddleback Church Was Too Flip," ABC News, September 7, 2008, http://abcnews.go.com/blogs/politics/2008/09/obama-my-answer (accessed September 19, 2014).

20. Lydia Saad, "Public Opinion about Abortion—An In-Depth Review," Gallup Poll, January 22, 2002, http://www.gallup.com/poll/9904/Public-Opinion-About-Abortion-InDepth-Review.aspx (accessed June 1, 2014).

21. "Widening Regional Divide over Abortion Laws," Pew Research Center for the People and the Press, July 29, 2013, http://www.peoplepress.org/2013/07/29/widening-regional-divide-over-abortion-laws/ (accessed June

1, 2014).

22. Tadeusz Pacholczyk, "Do Embryos Have Souls?" National Catholic Bio-ethics Center, March 2008, http://www.ncbcenter.org/page.aspx?pid=305 (accessed June 1, 2014).

23. Jon Henley, "'This Is Not a Life. I Can Lead It No More' Argument Rages across Europe in Wake of French Mercy Killing," *Guardian*, October 26, 2003, http://www.theguardian.com/world/2003/oct/27/health.france (accessed June 1, 2014).

24. Vincent Humbert, "Vincent Humbert's Letter to Chirac," BBC News, September 26, 2003, http://news.bbc.co.uk/2/hi/europe/3142366.stm (accessed September 19, 2014).

25. "God Alone Has the Right to Initiate and Terminate Life: Answers," *Christian Life Resources* http://www.christianliferesources.com/article/god-alone-has-the-right-to-initiate-and-terminate-life-answers-715 (accessed June 1, 2014).

26. "Euthanasia and Assisted Dying," BBC Religions, August 2009, http://www.bbc.co.uk/religion/religions/christianity/christianethics/euthanasia_1.shtml (accessed June 1, 2014).

27. "Views on End-of-Life Medical Treatments," Pew Research Religion and Public Life Project, November 21, 2013, http://www.pewforum.org/2013/11/21/views-on-end-of-life-medical-treatments/ (accessed June 1, 2014).

28. Epicurus, "Letter to Menoeceus," Epicurus & Epicurean Philosophy, http://www.epicurus.net/en/menoeceus.html (accessed September 19, 2014).

29. Woody Allen, *Without Feathers* (New York: Random House, 1975).

30. Arthur Schopenhauer, "On the Indestructibility of Our Essential Being by Death," in *The Indestructibility of Our Inner Nature*, trans. R. J. Hollingdale (New York: Viking, 1973).

31. Jean de La Fontaine, "The Astrologer Who Stumbled into a Well," in *The Fables of La Fontaine*, trans. Elizur Wright (Boston, 1841; University of Adelaide, 2014), https://ebooks.adelaide.edu.au/l/la_fontaine/jean_de/fables/complete.html#preface2.

32. Ecclesiastes 9:5.

33. Ecclesiastes 9:7–9.

34. Stephen Cave, *Immortality: The Quest to Live Forever and How It Drives Civilization* (New York: Crown, 2012), p. 266.

35. Steve Jobs, "Steve Jobs's 2005 Stanford Commencement Address," YouTube video, 15:04, http://www.youtube.com/watch?v=UF8uR6Z6KLc

(accessed June 1, 2014).

36. Christopher Hitchens, "Christopher Hitchens Closing Remarks," YouTube video, 2:41, https://www.youtube.com/watch?v=lhC459k1cMU (accessed September 19, 2014).

37. D'Souza, *Life after Death*, p. 3.

38. Ibid., pp. x–xi.

39. Steve Jobs, "Steve Jobs's 2005 Stanford Commencement Address."

BIBLIOGRAPHY

Alcock, James. "Back from the Future: Parapsychology and the Bem Affair." *Skeptical Inquirer*, March/April 2011. Available at Committee for Skeptical Inquiry. http://www.csicop.org/specialarticles/show/back_from_the_future.

Alexander, Eben. *Proof of Heaven: A Neurosurgeon's Journey into the Afterlife*. New York: Simon & Schuster, 2012.

Aristotle. *De Anima*. Oxford: Oxford University Press, 1979.

Baker, Mark C., and Stewart Goetz, eds. *The Soul Hypothesis: Investigations into the Existence of the Soul*. New York: Continuum, 2011.

Banerjee, K., O. Haque, and E. Spelke. "Melting Lizards and Crying Mailboxes: Children's Preferential Recall of Minimally Counterintuitive Concepts." *Cognitive Science* 37, no.7 (2013): 1251–89.

Barker, A. T., R. Jalinous, and I. L. Freeston. "Non-Invasive Magnetic Stimulation of the Human Motor Cortex." *Lancet* 8437 (1985): 1106–1107.

Bartsch, Karen and Henry Wellman. *Children Talk about the Mind*. Oxford: Oxford University Press, 1994.

Bayley, Barrington J. *The Seed of Evil*. London: Allison & Busby, 1979.

Beauregard, Mario, and Denyse O'Leary. *The Spiritual Brain: A Neuroscientist's Case for the Existence of the Soul*. New York: HarperOne, 2007.

Bering, Jesse, and David Bjorklund. "The Natural Emergence of Reasoning about the Afterlife as a Developmental Regularity." *Developmental Psychology* 40, no. 2 (2004): 217–33.

Biran, I., and A. Chatterjee. "Alien Hand Syndrome." *Archives of Neurology* 61 (2004): 292–94.

Blackmore, Susan. *Dying to Live: Science and the Near Death Experience*. London: Grafton, 1993.

Blanke, Olaf, Stéphanie Ortigue, Theodor Landis, and Margitta Seeck. "Stimulating Illusory Own-Body Perceptions." *Nature* 419 (2002): 269–70.

Bloom, Paul. *Descartes' Baby: How the Science of Child Development Explains What Makes Us Human*. New York: Basic Books, 2004.

———. "Religion, Morality, Evolution." *Annual Review of Psychology* 63 (2012): 179–99.

―――. "Worse Than Creationism: Evolution, Neuroscience, and the Responsibility of Psychologists." *American Psychological Society: Observer* 18, no. 10 (2005). http://www.psychologicalscience.org/index.php/publications/observer/2005/october-05/worse-than-creationism-evolution-neuroscience-and-the-responsibility-of-psychologists.html.

Boyer, Pascal. *Religion Explained: The Evolutionary Origin of Religious Thought.* New York: Basic Books, 2001.

Bremmer, Jan. *The Early Greek Concept of the Soul.* Princeton, NJ: Princeton University Press, 1983.

Brillouin, Léon. "Maxwell's Demon Cannot Operate: Information and Entropy I." *Journal of Applied Physics* 22 (1951): 334–37.

―――. *Science and Information Theory.* New York: Academic Press, 1956.

Casali, Adenauer, Olivia Gosseries, Mario Rosanova, Melanie Boly, Simone Sarasso, Karina R. Casali, Silvia Casarotto, et al. "A Theoretically Based Index of Consciousness Independent of Sensory Processing and Behavior." *Science Translational Medicine* 5, no.198 (2013): 105.

Cave, Stephen. *Immortality: The Quest to Live Forever and How It Drives Civilization.* New York: Crown, 2012.

―――. "What Science Says about the Soul." *Skeptic Magazine* 18, no. 2 (2013): 16–18.

Chomsky, Noam. *Language and Problems of Knowledge.* Cambridge, MA: MIT Press, 1988.

―――. "Linguistics and the Cognitive Science: Problems and Mysteries." In *The Chomskyan Turn*, edited by A. Kasher, pp. 26–53. Oxford: Basil Blackwell, 1991.

―――. "On the Nature, Use, and Acquisition of Language." In *Handbook of Child Language*, edited by William C. Ritchie and Tej K. Bhatia, pp. 33–54. New York: Academic Press, 1999.

―――. *The Prosperous Few and the Restless Many.* Berkeley, CA: Odonian, 1993.

Chopra, Deepak. *Life after Death: The Burden of Proof.* New York: Harmony Books, 2006.

Churchland, Patricia. "Moral Decision-Making and the Brain." In *Neuroethics: Defining the Issues in Theory, Practice, and Policy*, edited by Judy Illes, pp. 3–16. Oxford: Oxford University Press, 2006.

―――. *Touching a Nerve: The Self as Brain.* New York: W. W. Norton, 2013.

Churchland, Paul M. *Matter and Consciousness.* Cambridge, MA: MIT Press, 2013.

Clarke, Arthur C. "Hazards of Prophecy: The Failure of Imagination." Chapter 2 in

Profiles of the Future: An Inquiry into the Limits of the Possible. New York: Harper & Row, 1962.

Collins, Robin. "The Energy of the Soul." In *The Soul Hypothesis: Investigations into the Existence of the Soul,* edited by Mark Baker and Stewart Goetz, pp. 123–33. New York: Continuum, 2011.

Crick, Francis. *The Astonishing Hypothesis: The Scientific Search for the Soul.* New York: Charles Scribner's Sons, 1994.

Dawkins, Richard. *The God Delusion.* Boston: Houghton Mifflin, 2006.

Debruyne, H., M. Portzky, F. Van Den Eynde, and K. Audenaert. "Cotard's Syndrome: A Review." *Current Psychiatry Reports* 11 (2009): 197–202.

Dehaene, Stanislas. *Consciousness and the Brain: Deciphering How the Brain Codes Our Thoughts.* New York: Viking, 2014.

Dehaene, Stanislas, Lionel Naccache, Gurvan Le Clec'H, Etienne Koechlin, Michael Mueller, Ghislaine Dehaene-Lambert, Pierre-Francois van de Moortele, and Denis Le Bihan. "Imaging Unconscious Semantic Priming." *Nature* 395 (1998): 297–300.

Dennett, Daniel. *Consciousness Explained.* Boston: Little, Brown, 1991.

———. *Elbow Room: The Varieties of Free Will Worth Wanting.* Cambridge, MA: MIT Press, 1984.

———. *Freedom Evolves.* New York: Penguin, 2003.

———. *Intuition Pumps and Other Tools for Thinking.* New York: W. W. Norton, 2013.

———. *Sweet Dreams: Philosophical Obstacles to a Science of Consciousness.* Cambridge, MA: Bradford Books, 2006.

Desmurget, Michel, Karen T. Reilly, Nathalie Richard, Aleandru Szathmari, Carmine Mottolese, and Angela Sirigu. "Movement Intention after Parietal Cortex Stimulation in Humans." *Science* 324, no. 5928 (2009): 811–13.

D'Souza, Dinesh. *Life after Death: The Evidence.* Washington, DC: Regnery, 2009.

Ebbern, Hayden, Sean Mulligan, and Barry L. Beyerstein. "Maria's Near Death Experience: Waiting for the Other Shoe to Drop." *Skeptical Inquirer* 20, no. 4 (1996): 46–60.

Edgell, Penny, Joseph Gerteis, and Douglas Hartman. "Atheists as 'Other': Moral Boundaries and Cultural Membership in American Society." *American Sociological Review* 71 (2006): 211–34.

Einstein, Albert. *The World as I See It.* Secaucus, NJ: Citadel, 1999.

Elbert, Jerome W. *Are Souls Real?* Amherst, NY: Prometheus Books, 2000.

Feinberg, Todd E. *Altered Egos: How the Brain Creates the Self.* Oxford: Oxford University Press, 2001.

Flanagan, Owen. *The Problem of the Soul: Two Visions of the Mind and How to Reconcile Them*. New York: Basic Books, 2002.

Fodor, Jerry. "The Mind-Body Problem." In *The Mind-Body Problem*, edited by R. Warner and T. Szubka, pp. 24–40. Oxford: Blackwell, 1994.

Fox, Mark. *Religion, Spirituality, and the Near-Death Experience*. London: Routledge, 2002.

Freud, Sigmund. "Thoughts for the Times on War and Death." In vol. 14 of *The Standard Edition of the Complete Psychological Works of Sigmund Freud*, pp. 274–301. London: Hogarth, 1915.

Galak, Jeff, Leif D. Nelson, Robyn A. LeBoeuf, and Joseph Simmons. "Correcting the Past: Failures to Replicate Psi," *Journal of Personality and Social Psychology* 103, no. 6 (2012): 933–48.

Gallistel, Charles R. "Behaviorism, Methodological and Theoretical." In *Routledge Encyclopedia of Philosophy*, edited by Edward Craig, pp. 696–99. London: Routledge, 1998.

Gazzaniga, Michael S. "Cerebral Specialization and Interhemispheric Communication: Does the Corpus Callosum Enable the Human Condition?" *Brain* 123, no. 7 (2000): 1293–1326.

———. "Neurological Disorders and the Structure of Human Consciousness." *Trends in Cognitive Sciences* 7, no. 4 (2003): 161–65.

Goetz, Stewart, and Charles Taliaferro. *A Brief History of the Soul*. Hoboken, NJ: Wiley-Blackwell, 2011.

Grayling, A. C. *Descartes: The Life and Times of a Genius*. New York: Walker, 2006.

Green, Joel, ed. *What about the Soul? Neuroscience and Christian Anthropology*. Nashville: Abingdon, 2004.

Greene, Joshua D. *Moral Tribes: Emotion, Reason, and the Gap between Us and Them*. New York: Penguin, 2013.

———. "Social Neuroscience and the Soul's Last Stand." In *Social Neuroscience: Toward Understanding the Underpinnings of the Social Mind*, edited by A. Todorov, S. Fiske, and D. Prentice, pp. 263–73. New York: Oxford University Press, 2011.

Greene, Joshua, and Jonathan Cohen. "For the Law, Neuroscience Changes Nothing and Everything." *Philosophical Transactions of the Royal Society of London* 359 (2004): 1775–85.

Haxby, J. V. "Distributed and Overlapping Representations of Faces and Objects in Ventral Temporal Cortex." *Science* 293 (2009): 2425–30.

Haynes, J. D., and R. Geraint. "Decoding Mental States from Brain Activity in Humans." *Nature Neuroscience* 7 (2006): 523–34.

Hitchens, Christopher. *God Is Not Great: How Religion Poisons Everything*. New York: Twelve Books, 2007.

————. *The Portable Atheist: Essential Readings for the Non-Believer*. Boston: Da Capo, 2007.

Hoffman, Donald D., and Manish Singh. "Computational Evolutionary Perception." *Perception* 41 (2012): 1073–91.

Hoffman, Peter. *Life's Ratchet: How Molecular Machines Extract Order from Chaos*. New York: Basic Books, 2012.

Hofstadter, Douglas. *I Am a Strange Loop*. New York: Basic Books, 2008.

Holden, Janice, Bruce Greyson, and Debbie James. *The Handbook of Near-Death Experience: Thirty Years of Investigation*. Santa Barbara, CA: Praeger, 2009.

Humbert, Vincent. *Je Vous Demande le Droit de Mourir*. Neuilly-sur-Seine, France: Michel Lafon, 2003.

Hume, David. *An Enquiry Concerning Human Understanding*. 1748. Indianapolis: Hackett, 1993.

Kaku, Michio. *Physics of the Impossible: A Scientific Exploration into the World of Phasers, Force Fields, Teleportation, and Time Travel*. New York: Anchor Books, 2009.

Kelly, Edward, and Emily Kelly. *Irreducible Mind: Toward a Psychology for the 21st Century*. Lanham, MD: Rowman & Littlefield, 2009.

Kim, Jaegwon. *Physicalism or Something Near Enough*. Princeton, NJ: Princeton University Press, 2007.

Kuhlmeier, Valerie, Paul Bloom, and Karen Wynn. "Do 5-Month-Old Infants See Humans as Material Objects?" *Cognition* 94 (2004): 95–103.

Long, Jeffrey, and Paul Perry. *Evidence of the Afterlife: The Science of Near-Death Experiences*. New York: HarperCollins, 2010.

MacDougall, Duncan. "Hypothesis Concerning Soul-Substance Together with Experimental Evidence of the Existence of Such Substance." *American Medicine* (April 1907): 240–43.

Mauro, James. "Bright Lights, Big Mystery." *Psychology Today* (July 1992). http://www.psychologytoday.com/articles/200910/bright-lights-big-mystery.

Maxwell, James. *Theory of Heat*. Rev. edition, edited by John William Strutt Rayleigh. London: Longmans, Green, 1902.

Moody, Raymond. *Life after Life*. Atlanta, GA: Mockingbird Books, 1975.

Morowitz, Harold. "The Mind Body Problem and the Second Law of Thermodynamics." *Biology and Philosophy* 2 (1987): 271–75.

Nahmias, Eddy, D. Justin Coates, and Trevor Kvaran. "Free Will, Moral Responsibility, and Mechanism: Experiments on Folk Intuitions." *Midwest Studies in Philosophy* 31 (2007): 214–42.

Nichols, Shaun. "The Folk Psychology of Free Will: Fits and Starts." *Mind & Language* 19, no. 5 (2004): 473–502.

Norris, Pippa, and Ronald Inglehart. *Sacred and Secular: Religion and Politics Worldwide.* New York: Cambridge University Press, 2004.

Onishi, Kristine, and Renée Baillargeon. "Do 15-Month-Old Infants Understand False Beliefs?" *Science* 308, no. 5719 (2005): 255–58.

Paul, Gregory S. "Cross-National Correlations of Quantifiable Social Health with Popular Religiosity and Secularism in the Prosperous Democracies: A First Look." *Journal of Religion & Society* 7 (2005): 1–17.

Penfield, Wilder. *The Mystery of the Mind: A Critical Study of Consciousness and the Human Brain.* Princeton, NJ: Princeton University Press, 1978.

Pfungst, Oskar. *Clever Hans (The Horse of Mr. Von Osten): A Contribution to Experimental Animal and Human Psychology.* New York: Henry Holt, 1911.

Pinker, Steven. *The Blank Slate: The Modern Denial of Human Nature.* New York: Penguin Books, 2003.

——. *How the Mind Works.* New York: Norton, 1999.

Plato. *Phaedo.* Oxford: Oxford University Press, 2009.

——. *The Republic.* New York: Penguin Classics, 2007.

Putnam, Robert, and David Campbell. *American Grace: How Religion Divides and Unites Us.* New York: Simon & Schuster, 2012.

Ring, Kenneth, and Sharon Cooper. *Mindsight: Near-Death and Out-of-Body Experiences in the Blind.* Palo Alto, CA: William James Center for Consciousness Studies, Institute of Transpersonal Psychology, 1999.

Russell, Bertrand. "Is There a God?" (1952). In *The Collected Papers of Bertrand Russell,* vol. 11, *Last Philosophical Testament, 1943–68,* edited by John G. Slater and Peter Köllner, pp. 543–48. London: Routledge, 1997. http://www.personal.kent.edu/~rmuhamma/Philosophy/RBwritings/isThereGod.htm.

——. *Religion and Science.* Oxford: Oxford University Press, 1997.

Ryle, Gilbert. *The Concept of Mind.* Chicago: University of Chicago Press, 1949.

Sagan, Carl. *The Demon-Haunted World: Science as a Candle in the Dark.* New York: Random House, 1995.

Schachter, Stanley, and Jerome E. Singer. "Cognitive, Social, and Physiological Determinants of Emotional State." *Psychological Review* 69 (1962): 379–99.

Schlosshauer, Maximilian, Johannes Koffler, and Anton Zeilinger. "The Interpretation of Quantum Mechanics: From Disagreement to Consensus?" *Annalen der Physik* 525, no. 4 (2013): A51–A54.

Shermer, Michael. *The Science of Good and Evil: Why People Cheat, Gossip, Care, Share, and Follow the Golden Rule.* New York: Times Books, 2004.

Shwed, Uri, and Peter Bearman. "The Temporal Structure of Scientific Consensus Formation." *American Sociological Review* 75, no. 6 (2010): 817–40.

Simmons, Joseph P., Leif D. Nelson, and Uri Simonsohn. "False-Positive Psychology: Undisclosed Flexibility in Data Collection and Analysis Allows Presenting Anything as Significant." *Psychological Science* 22, no. 11 (2011): 1359–66.

Simpson, S. M. "Near Death Experience: A Concept Analysis as Applied to Nursing." *Journal of Advanced Nursing* 36, no. 4 (2001): 520–26.

Slater, Alan, Victoria Morison, and David Rose. "Habituation in the Newborn." *Infant Behavior and Development* 7 (1984): 183–200.

Smith, Christopher. "Julien Offray de La Mettrie (1709–1751)." *Journal of the History of the Neurosciences* 11, no. 2 (2002): 110–24.

Snobelen, Stephen D. "Isaac Newton, Heretic: The Strategies of a Nicodemite." *British Journal for the History of Science* 32, no. 4 (1999): 381–419.

Spelke, Elizabeth, R. Kestenbaum, D. J. Simons, and D. Wein. "Spatiotemporal Continuity, Smoothness of Motion and Object Identity in Infancy." *British Journal of Developmental Psychology* 13, no. 2 (1995): 113–42.

Stenger, Victor J. *God: The Failed Hypothesis; How Science Shows That God Does Not Exist.* Amherst, NY: Prometheus Books, 2007.

————. *God and the Folly of Faith: The Incompatibility of Science and Religion.* Amherst, NY: Prometheus Books, 2012.

————. *The New Atheism: Taking a Stand for Science and Reason.* Amherst, NY: Prometheus Books, 2009.

————. *Physics and Psychics: The Search for a World beyond the Senses.* Amherst, NY: Prometheus Books, 1990.

————. *Quantum Gods: Creation, Chaos, and the Search for Cosmic Consciousness.* Amherst, NY: Prometheus Books, 2009.

————. *The Unconscious Quantum: Metaphysics in Modern Physics and Cosmology.* Amherst, NY: Prometheus Books, 1995.

Tonry, Michael. *Thinking about Crime: Sense and Sensibility in American Penal Culture.* Oxford: Oxford University Press, 2004.

Tyson, Neil deGrasse. "The Perimeter of Ignorance: A Boundary Where Scientists Face a Choice: Invoke a Deity or Continue the Quest for Knowledge." *Natural History Magazine* 28, no. 5 (November 2005). http://www.natural historymag.com/universe/211420/the-perimeter-of-ignorance.

————. *The Sky Is Not the Limit: Adventures of an Urban Astrophysicist.* Amherst, NY: Prometheus Books, 2004.

Wassermann, E., and J. Grafman. "Combining Transcranial Magnetic Stimulation and Neuroimaging to Map the Brain." *Trends in Cognitive Sciences* 1, no. 6 (1997): 199–200.

Woerlee, Gerald. *Mortal Minds: The Biology of Near Death Experiences.* Amherst, NY: Prometheus Books, 2005.

Woodward, Amanda. "Infants Selectively Encode the Goal Object of an Actor's Reach." *Cognition* 69, no.1 (1998): 1–34.

Young, Liane, Joan Albert Camprodon, Marc Hauser, Alvaro Pascual-Leone, and Rebecca Saxe. "Disruption of the Right Temporoparietal Junction with Transcranial Magnetic Stimulation Reduces the Role of Beliefs in Moral Judgments." *Proceedings of the National Academy of Sciences of the United States of America* 107, no.15 (2010): 6753–58.

Zimbardo, Phil, and John Boyd. *The Time Paradox: The New Psychology of Time That Will Change Your Life.* New York: Atria Books, 2008.

Zuckerman, Phil. *Society without God: What the Least Religious Nations Can Tell Us about Contentment.* New York: New York University Press, 2008.

INDEX